AF871033

Springer Series in Synergetics

Editor-in-Chief

Peter Schuster, Theoretical Chemistry and Structural Biology, University of Vienna, Vienna, Austria

Series Editors

Henry D.I. Abarbanel, Institute for Nonlinear Science, University of California, San Diego, USA

Dan Braha, New England Complex Systems Institute, Cambridge, USA

Péter Érdi, Center for Complex Systems Studies, Kalamazoo College, USA, Hungarian Academy of Sciences, Budapest, Hungary

Karl J Friston, Institute of Cognitive Neuroscience, University College London, London, UK

Hermann Haken, Center of Synergetics, University of Stuttgart, Stuttgart, Germany

Viktor Jirsa, Centre National de la Recherche Scientifique (CNRS), Université de la Méditerranée, Marseille, France

Janusz Kacprzyk, Systems Research, Polish Academy of Sciences, Warsaw, Poland

Jürgen Kurths, Nonlinear Dynamics Group, University of Potsdam, Potsdam, Germany

Ronaldo Menezes, Computer Science Department, University of Exeter, Exeter, FL, USA

Andrzej Nowak, Department of Psychology, Warsaw University, Warsaw, Poland

Hassan Qudrat-Ullah, Decision Sciences, York University, Toronto, Canada

Linda Reichl, Center for Complex Quantum Systems, University of Texas, Austin, USA

Frank Schweitzer, System Design, ETH Zurich, Zurich, Switzerland

Didier Sornette, Entrepreneurial Risk, ETH Zurich, Zurich, Switzerland

Stefan Thurner, Section for Science of Complex Systems, Medical University of Vienna, Vienna, Austria

Springer Series in Synergetics

Founding Editor: H. Haken

The Springer Series in Synergetics was founded by Herman Haken in 1977. Since then, the series has evolved into a substantial reference library for the quantitative, theoretical and methodological foundations of the science of complex systems.

Through many enduring classic texts, such as Haken's *Synergetics and Information and Self-Organization*, Gardiner's *Handbook of Stochastic Methods*, Risken's *The Fokker Planck-Equation* or Haake's *Quantum Signatures of Chaos*, the series has made, and continues to make, important contributions to shaping the foundations of the field.

The series publishes monographs and graduate-level textbooks of broad and general interest, with a pronounced emphasis on the physico-mathematical approach.

Vladimir Semenov

Electronic Modelling of Deterministic and Stochastic Oscillators

Experimental Implementations of Ordinary Differential Equations

Springer

Vladimir Semenov
Computer Modelling in Electronics
Saratov State University
Saratov, Russia

ISSN 0172-7389 ISSN 2198-333X (electronic)
Springer Series in Synergetics
ISBN 978-3-031-81552-2 ISBN 978-3-031-81553-9 (eBook)
https://doi.org/10.1007/978-3-031-81553-9

© The Editor(s) (if applicable) and The Author(s), under exclusive license to Springer Nature Switzerland AG 2024

This work is subject to copyright. All rights are solely and exclusively licensed by the Publisher, whether the whole or part of the material is concerned, specifically the rights of translation, reprinting, reuse of illustrations, recitation, broadcasting, reproduction on microfilms or in any other physical way, and transmission or information storage and retrieval, electronic adaptation, computer software, or by similar or dissimilar methodology now known or hereafter developed.
The use of general descriptive names, registered names, trademarks, service marks, etc. in this publication does not imply, even in the absence of a specific statement, that such names are exempt from the relevant protective laws and regulations and therefore free for general use.
The publisher, the authors and the editors are safe to assume that the advice and information in this book are believed to be true and accurate at the date of publication. Neither the publisher nor the authors or the editors give a warranty, expressed or implied, with respect to the material contained herein or for any errors or omissions that may have been made. The publisher remains neutral with regard to jurisdictional claims in published maps and institutional affiliations.

This Springer imprint is published by the registered company Springer Nature Switzerland AG
The registered company address is: Gewerbestrasse 11, 6330 Cham, Switzerland

If disposing of this product, please recycle the paper.

To my family

Preface

This book deals with methods of electronic modelling of dynamical systems described by ordinary differential equations. During the process of writing, I was guided by two purposes. The first one is to provide a set of tools and ready-made solutions to specialists in nonlinear dynamics who need experimental realization of studied mathematical models. I hope that this would enable researchers who are not focused on problems of electronics and circuit theory to implement their ideas in real devices and to obtain experimental confirmation of numerical and theoretical results in short time. For this reason the book contains additional sections dedicated to functioning of an operational amplifier and related circuits, development of noise sources, problems of measurements, etc. The second aim is demonstration for specialists in electronic engineering that a wide variety of non-electronic systems and their intrinsic peculiarities can be realized by using electronic circuits. Generally, the book motto is 'Electronics for theorists and mathematics, nonlinear dynamics for experimenters and electronic engineers'.

The term 'dynamical system' and related topics are briefly described in the first chapter before the issues of electronic realization of dynamical systems begin to be considered. Of course, general description of the problematics is not deep. However, it can be helpful for quick review and introduction to the terminology. This chapter may be used as an additional 'dictionary' for students and researchers who do not directly work on problems of nonlinear dynamics.

The second and third chapters are fully devoted to issues of circuit theory and electronics. First, basic relations and laws are considered. After that the problematics is continuously changed to questions being relevant to electronic modelling of dynamical systems. Besides description and generalization of well-known electronic modelling methods, the chapters include several technics and particular circuits developed by the author.

Particular and very specific issues are combined into the fourth chapter. For instance, realization of time-delay and development of noise sources are among them. These problems are less frequent, but their solutions require a lot of time and imply unusual approaches.

The fifth chapter represents a symbiosis of nonlinear dynamics and electronics. On the one hand, it includes examples of electronic implementation of different dynamical systems by means of classical approaches and analog modelling. On the other hand, various deterministic and stochastic effects and the corresponding bifurcation phenomena briefly discussed in the first chapter are described and explained on examples of the mentioned dynamical systems.

I hope this book would be interesting for a large circle of readers. It may be also expected to save scientists' time and energy spent on purely practical details. It allows to concentrate on ideas and new possible fields of activity.

All the analog models presented in the book were developed by the author in the frameworks of experimental explorations of nonlinear phenomena. For the convenience of readers these circuits are described in details with concrete nominal values of all elements. Full description of computer programs used for realization of time-delay lines and sources of noise are also included.

I am very grateful to my family and to all persons who exerted positive influence during different periods of my life for their support and patience: Tatiana Vadivasova, Vladimir Astakhov, Galina Strelkova, Artur Khokhlov, Alexander Neiman, Vladimir Malyaev, Sergey Koblyansky, Anna Zakharova, Yuri Maistrenko, Daniel Brunner, Javier Porte and Laurent Larger.

In addition, I acknowledge support from the Russian Science Foundation under Grant No. 24-72-00054.

Saratov, Russia Vladimir Semenov

Synthesis of Dynamical Systems and This Book

Similarly to other fields of science, nonlinear dynamics is based on theoretical and experimental approaches going hand in hand. Theoretical knowledges in nonlinear dynamics are results of applied methods of abstractions, idealizations and hypotheses, clarifications of mathematical models. The main goal of such activity is formulation of general laws and properties of phenomena observed in different scientific areas. It is impossible to obtain theoretical knowledges without accumulation and generalization of empirical materials. On the other hand, description of experimental data which does not involve further theoretical understanding is only a statement of observation. It does not reveal nature of phenomena and does not allow to compare obtained results with existing scientific perceptions and concepts. Then a comparative analysis of theoretical and experimental data is essentially important for study of processes and systems of any nature.

Two approaches of experimental exploration can be distinguished. The first way consists in study of already existing objects (planet Earth, a pack of animals, a backward-wave tube, atmospheric layers, etc.). The second one implies synthesis of a studied system with its own intrinsic peculiarities for further study. In such a case one carries out modelling of the researched object.

Generally, one can divide modelling methods into a numerical simulation and physical modelling. The numerical simulation aims to find an approximate solution of mathematical model equations corresponding to certain initial conditions. Modelling in a physical experiment involves creation of a prototype of the studied model as a physical object. In other words, it is assumed that the dynamics of the created object obeys the mathematical model equations. Methods of electronic modelling of oscillatory dynamical systems became a frequent practice. In the frameworks of that methods one creates electronic circuits whose dynamics is described by equations of the studied mathematical model.

Flurry of activity in the area of electronic modelling arose in the last decade of the twentieth century. On the one hand, it happened because of increased abilities of electronic microchips and integrated circuits. At the same time, the computation power of existing computers could not satisfy all demands from researchers. As a result, in certain cases electronic modelling was the only appropriate approach to experimentally check the correctness of theoretical hypotheses. The reader can ask the reasonable question: 'Why electronic modelling is still actual today, when one can simulate studied models numerically and check its correctness without any significant problems?' There are several reasons. First of all, analytical methods of solution derivations are approximate technics as well as methods of numerical integration of model equations. Such approach can give rise to wrong results. Among such situations is studying dynamical systems close to bifurcations, where the systems are structural-unstable, and exploration of stochastic dynamical systems in the presence of high-amplitude noise. Moreover, it is well-known that numerical integration methods can accumulate computational inaccuracies. Anyway, importance of a physical experiment, which can be carried out by means of electronic modelling, is obvious regardless the specifics of a researched problem. This is due to the fact that experimental confirmation of theoretical and numerical results is an objective criterion of their reliability.

The current book is an attempt of the author to systematize the methods of electronic modelling and to present them in the same manner. All approaches of electronic modelling can be conditionally divided into two kinds: a classical experiment on oscillatory circuits and analog modelling. Both of them are considered in the book.

Contents

Chapter 1
Continuous Dynamical System

The first chapter introduces the reader to basic terms and concepts associated with dynamical systems before proceeding with description of electronic modelling methods. First of all, it is necessary to correctly introduce the main term 'Dynamical system'. All necessary definitions are brought in the first chapter. Issues of classification of the dynamical systems and description methods are considered. The first chapter has an overview character. If more detailed study is necessary, the reader can explore the discussed topics in pages of books where the material is more rigorous and deep (see for instance books [1–4]). Probably, the first chapter is less interesting for experts in nonlinear dynamics. However, it may be useful for specialists in electronic engineering.

1.1 Mathematical Model of a Dynamical System

A dynamical system (DS) is any object or process evolving in accordance with a certain law and described by a set of quantities uniquely identifying its state at a given time moment. The evolution law is a functional dependence, which describes the evolution of an initial state over time. Initially, the term 'Dynamical system' was introduced in the context of mechanical processes described by Newton's dynamics laws. Nowadays the definition of the DS involves systems and processes of any nature (physical, biological, chemical, geological, climatic, economical, social, informational, etc.). Depending on the specifics of the evolution law, the DS can be defined in different manner. For example, the evolution law (evolution operator) can be determined using differential equations or discrete maps. The choice of a description approach establishes a concrete kind of a DS model. **The mathematical model of a dynamical system** is an image of an object or process presented from the mathematical point of view. A mathematical model state is uniquely defined by **dynamical variables** (coordinates) and a specified evolution law.

© The Author(s), under exclusive license to Springer Nature Switzerland AG 2024

V. Semenov, *Electronic Modelling of Deterministic and Stochastic Oscillators*,
Springer Series in Synergetics, https://doi.org/10.1007/978-3-031-81553-9_1

Different mathematical models can be used for description of the same DS. Usage of a concrete mathematical model is determined by the degree of approximation and objective conditions. Development of the mathematical model is often accompanied by finding a balance between simplicity and universality. A mathematical model exhibiting several peculiarities of a considered system can be overly complicated. In such a case, it may be more rational to develop several simplified mathematical models reflecting certain properties of the system. A good example of such approach is a broad manifold of models of neurons and synapses [5], which demonstrate different kinds of the single and collective behaviour observed in neural systems of living things. Such effects are excitability, the chaotic dynamics, self-excitation, synchronization, propagation of neural impulses, etc.

Other examples can be mentioned. Let us consider motion of a point mass, which is suspended from a string (Fig. 1.1). A simple mathematical model of this system needs to be developed. It requires exclusion of some factors from the consideration. The weight is assumed to be small (distribution of mass in space is neglected) and suspended from a massless inextensible cord, which always remains to be taut. The motion does not lose energy to friction or air resistance. The pendulum is considered in an inertial reference system, where the pendulum pivot rests or moves uniformly and rectilinearly. The body mass m suspended from the inextensible cord of length l is driven by the gravity $\vec{F_1} = m\vec{g}$ and a restoring force $\vec{F_2}$. Then Newton's second law can be written: $m\vec{a} = m\vec{g} + \vec{F_2}$. An equation for tangent projections of acceleration and force vectors becomes: $ma_\tau = -mg\sin(\varphi)$, where φ is a deflection angle. The sign minus means that a tangent gravity component is opposite to a pendulum displacement. The tangent acceleration component a_τ is related with the deflection angle φ by the formula $a_\tau = l(d^2\varphi/dt^2)$. Then an equation for the pendulum oscillations has the following form:

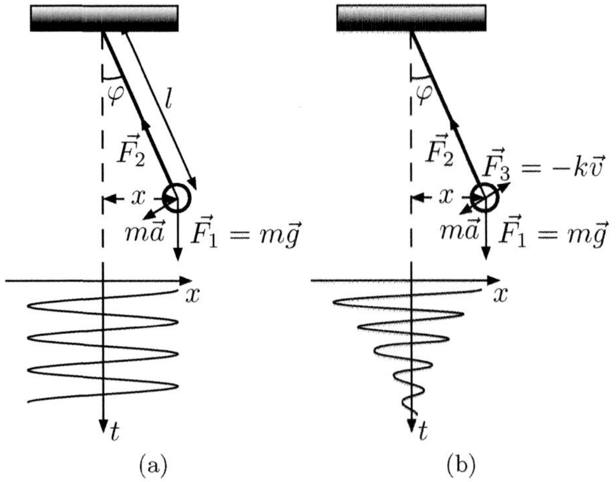

Fig. 1.1 Forces acting on a mass point suspended from a string and the evolution of a displacement $x(t)$ from an equilibrium state: (**a**) idealized simple pendulum; (**b**) model with friction

$$\frac{d^2\varphi}{dt^2} + \omega_0^2 \sin(\varphi) = 0, \tag{1.1}$$

where $\omega_0 = \sqrt{g/l}$. In a case of a small deflection angle one can apply the approximation $\sin(\varphi) \approx \varphi$ (linear approximation) and transform the equation of motion:

$$\frac{d^2\varphi}{dt^2} + \omega_0^2\varphi = 0. \tag{1.2}$$

Equations (1.2) and (1.1) describe the same system but have the principal difference—the presence of nonlinearity. Equation (1.2) is linear but it can be used only for a case of a small deflection angle. Equation (1.1) is useful for a larger range of values $\varphi(t)$, although it is nonlinear and more difficult for analysis. The amplitude of oscillations $\varphi(t)$ dictates the possibility to apply either Eq. (1.2) or (1.1). In turn, a scale of the oscillations $\varphi(t)$ depends on values of an initial displacement and an initial velocity (**initial conditions**). A displacement $x(t)$ of the point mass from a pivot coordinate also can be chosen as a dynamical variable. Using a linear approximation in a case of small angles of deflection, $x = l\sin(\varphi) \approx l\varphi$, one can rewrite the equation of motion in terms of the variable x:

$$\frac{d^2x}{dt^2} + \omega_0^2 x = 0. \tag{1.3}$$

Equations (1.2) and (1.3) are models of a harmonic oscillator whose solution has the form $A\cos(\omega_0 t + \varphi_0)$. Particularly, Eq. (1.3) has the solution $x(t) = x_{max}\cos(\omega_0 t + \varphi_0)$ (see Fig. 1.1a), where x_{max} and φ_0 are constants defined by initial conditions.

Mathematical models (1.2) and (1.1) do not take into consideration the action of friction. Friction at the pivot point and air resistance inevitably present in real pendulums. Let us consider how the air resistance impacts on the pendulum dynamics (Fig. 1.1b). In such a case an additional force of friction, $\vec{F_3}$, acts in the opposite direction to a velocity vector $\vec{v}$. The force of friction is assumed to be proportional to the velocity: $\vec{F_3} = -k\vec{v}$. A numerical value of the velocity is $v = l(d\varphi/dt)$. Using the condition of small amplitudes, one can derive an equation for tangent projections of the forces and acceleration $ma_\tau = -mg\varphi - kl(d\varphi/dt)$. Then the pendulum equation becomes:

$$\frac{d^2\varphi}{dt^2} + \frac{k}{m}\frac{d\varphi}{dt} + \frac{g}{l}\varphi = 0. \tag{1.4}$$

A further step is to denote $\omega_0 = \sqrt{g/l}$, $2\delta = k/m$ and to rewrite Eq. (1.4) for the variable x:

$$\frac{d^2x}{dt^2} + 2\delta\frac{dx}{dt} + \omega_0^2 x = 0. \tag{1.5}$$

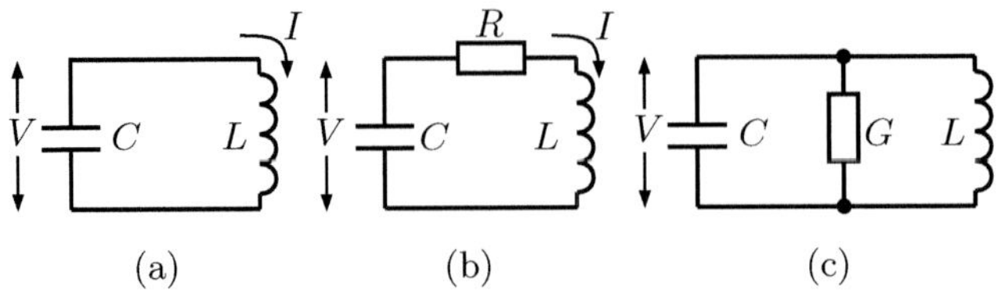

Fig. 1.2 Linear oscillatory circuit: (**a**) model of a LC-circuit without resistance; (**b**) series oscillatory LCR-circuit; (**c**) parallel oscillatory LCR-circuit

Model (1.5) exhibits periodic damped oscillations in the presence of weak friction (Fig. 1.1b). The corresponding solution is $x(t) = x_{max} \exp(-\delta t) \cos(\omega_0 t + \varphi_0)$. System (1.5) is called a **linear oscillator with dissipation**.

Another basic example of an oscillator is an oscillatory circuit. Firstly, an ideal circuit without resistance should be considered. It is made up of a capacitor and an inductor (Fig. 1.2a). The ideal capacitor model is characterized by the linear dependence of a charge q on the capacitor plates on an applied voltage V: $q = CV$. A current passing through the capacitor is a time derivative of the charge. It allows to write $I = C(dV/dt)$. A perfect inductor model involves a linear relationship between a magnetic flux linkage and a passing current: $\Phi = LI$. Then a relationship between a time derivative of the passing current and the applied voltage becomes: $L(dI/dt) = V$. According to Kirchhoff's second law, the algebraic sum of voltage drops around a loop equals to the summary electromotive force. Since the circuit depicted in Fig. 1.2a does not include any sources of the electromotive force, one can write $V_L + V_C = 0$ (V_L is a voltage across the inductor, V_C is a voltage across the capacitor). The same current I passes through the inductor and the capacitor. Then Kirchhoff's second law becomes $L(dI/dt) + V_C = 0$ or

$$\frac{d^2V}{dt^2} + \omega_0^2 V = 0, \tag{1.6}$$

where $V = V_C$ and $\omega_0 = 1/\sqrt{LC}$.

It is evident that the difference between Eqs. (1.2), (1.3) and (1.6) consists is the dimension of the dynamical variables φ, V and x. These equations are mathematically equivalent and have the same dimensionless form. Transition to the dimensionless form is carried out by using the substitution $y = x/x_0$, $y = V/V_0$ and $y = \varphi$, where $x_0 = 1$ meter and $V_0 = 1$ V. A dimensionless time variable is introduced as being: $t_* = \omega_0 t$. As a result, the variable y and the time t_* are dimensionless and the equations have the dimensionless form $\ddot{y} + y = 0$, where $\ddot{y} = d^2y/dt_*^2$.

Model (1.6) was derived using the following assumptions: validity of the linear relationships $\Phi = LI$ and $q = CV$; zero resistance of wires and an inductor; the absence of a leakage current through a capacitor. Let us study a circuit with non-zero resistance (Fig. 1.2b). The summary resistance R of all circuit elements is schematically united into one resistor. Kirchhoff's second law for the circuit in Fig. 1.2b becomes: $V_L + V_R + V_C = 0$. Using Ohm's law, the following equation can be derived:

$$\frac{d^2V}{dt^2} + 2\delta\frac{dV}{dt} + \omega_0^2 V = 0, \quad (1.7)$$

where $\omega_0 = 1/\sqrt{LC}$ and $2\delta = R/L$. Equations (1.5) and (1.7) are equivalent. A mathematical model of a parallel oscillatory LCR-circuit (Fig. 1.2c) is also transformed into the form (1.7) with $\omega_0 = 1/\sqrt{LC}$, $2\delta = G/C$ (G is the resistor conductance).

Generality of the phenomena described above on examples of the mechanical and electrical systems is evident. Analyzing the same mathematical model, we can simultaneously study a manifold of effects of different nature. Depending upon a stated objective, a model can be conservative or dissipative, linear or nonlinear, etc. The models mentioned above have different properties and belong to different classes of oscillators. The issues of classification are discussed in the further section.

1.2 Classification of Dynamical Systems

A dynamical system[1] is defined if the following elements are given:

- A set of states X forming system's phase space;
- A set of time moments Q;
- An evolution operator $F_{t_0}^{\tau}$, which unambiguously transforms each state $\vec{x}_0 \in X$ at the initial time moment $t_0 \in Q$ into a state $\vec{x}_t \in X$ at another time $t = t_0 + \tau \in Q$.

A classification of DSs is based on the properties of the evolution operator and the sets X and Q. If Q is a continuous manifold of real numbers $Q \in R$, i.e. time is given by a continuous set of values, and the evolution operator is continuous, then the DS is called a **continuous system** or a **continuous-time system** or a **flow**. If the set Q is countable, then the DS is called a **discrete-time system** or a **cascade** or an **iterated map**. Systems (1.1)–(1.7) belong to a class of continuous-time systems, since $t \in R$. For instance, the DSs with a countable manifold of time moments describe climate changes or changes in animal populations registered with a certain period T. Then a functional dependence has the form $x(t_0 + (n+1)T) = f(x(t_0 + nT))$, where t_0 is the initial moment of observation.

Similarly to the set of time moments Q, the set of states X can be continuous or countable (finite). Typical examples of DSs with the countable sets of states are **cellular automata**. DSs given by the continuous sets of time and states of dimension N ($X \in \mathbb{R}^N$) are systems of ordinary differential equations. The set X can be a

[1] In the following, the term 'dynamical system' is used to refer to its mathematical model.

functional space. In such a case the DS represents a system described by partial differential equations, integro-differential equations, ordinary equations with time delay. Systems with a finite number of degrees of freedom are called **lumped**. Their phase space is finite-dimensional. If the system state is described by a function of spatial coordinates, then the system is called a **distributed system** or a **medium**. If spatial coordinates are defined on a continuous set, the number of degrees of freedom becomes infinite. It means an infinite number of quantities is required to define the system state.

The evolution operator can have different properties, which allows to classify the DSs. For example, **linear** and **nonlinear** DSs are distinguished. The evolution operator of the linear DS satisfies the superposition principle: $F_t^\tau(\vec{a}+\vec{b}) = F_t^\tau(\vec{a}) + F_t^\tau(\vec{b})$. If the evolution operator does not satisfy the superposition principle then the DS is nonlinear. If the evolution operator is defined for both positive and negative time values τ, the DS is reversible. If the operator is defined only for $\tau \geq 0$, the DS is irreversible in time.

If the evolution operator does not include a functional dependence on a time moment, the DS is called **autonomous**. Otherwise the DS is **nonautonomous**. Models (1.1)–(1.7) belong to a class of autonomous systems. In terms of physics, the presence of an explicit dependence on time means the action of an external influence or time-varying parameters. An example of a nonautonomous DS is a LCR-oscillatory circuit including the periodic electromotive force $V_{\text{ext}} = V_0 \sin(\omega_{\text{ext}} t)$.

If the evolution operator saves the phase volume then the DS is called **conservative**. In terms of physics it means that the full energy of the conservative DS is constant. For example, the full mechanical energy being the sum of potential and kinetic energies is constant in pendulum models (1.1) and (1.2) without friction. Similarly, the total energy of a LC-circuit (see Eq. (1.6)), which is the sum of electric field energy stored in a capacitor and energy of magnetic field induced in an inductor, is constant. Systems losing energy due to friction or dissipation are **dissipative** (pendulum model (1.4) with air resistance or LCR-circuit model (1.7)).

Oscillatory systems (or **oscillators**) represent a special group of the DS. These systems can generate oscillations. The oscillatory systems are divided into classes. Depending on properties, one distinguishes oscillatory systems being linear or nonlinear, lumped or distributed, conservative or dissipative, autonomous or nonautonomous. **Oscillatory systems with lumped parameters** are described by a finite number of ordinary differential equations. The phase space of such systems is finite-dimensional. **Oscillatory systems with distributed parameters** are described by partial differential equations or by equations with time delay and have an infinite number of degrees of freedom.

An especially important group of oscillatory systems is made up of **self-oscillators** (or **self-sustained oscillators**) being principally nonlinear and dissipative. They have an internal source of energy providing the existence of nondecaying oscillations (**self-oscillations** or **self-sustained oscillations**). Characteristics of such oscillations (amplitude, frequency, waveform, etc.) are dictated by system parameters and do not depend on initial conditions in a certain area of the phase space.

Finally, one divides DSs into deterministic and stochastic ones. The state of the **deterministic DS** at any time is unambiguously determined by initial conditions and the evolution operator. The evolution operator of the **stochastic DS** includes a random component and does not allow to unambiguously define the system state. The stochastic DSs require a probabilistic description. From the viewpoint of physics, the presence of the random component is associated with the existence of internal noise sources or external random perturbations.

1.3 Phase Space and a Phase Portrait

Let us consider systems with continuous sets of time Q and states X modelled by a finite number of ordinary differential equations. Such systems can be described in terms of mechanical terminology. In the frameworks of this approach one introduces a set of quantities $x_1, x_2, \ldots, x_N$ describing the system and their values at the initial time moment t_0. The initial values of the quantities x_i are arbitrary, but two different sets x_i and x_i^* correspond to different states of the system. A law of evolution of the system state in time is given by a system of ordinary differential equations:

$$\frac{dx_i}{dt} = \dot{x}_i = f_i(x_1, x_2, \ldots, x_N), \quad i = 1, 2, \ldots, N. \tag{1.8}$$

System (1.8) can be written in a vectorial form:

$$\dot{\vec{x}} = \vec{F}(\vec{x}), \tag{1.9}$$

where $\dot{\vec{x}} = d\vec{x}/dt$, i.e. $\vec{F}(\vec{x})$—is a vectorial N-dimensional function.

Consideration of the quantities $x_1, x_2, \ldots, x_N$ as coordinates of a point $\vec{x}$ in a N-dimensional space allows to get an evident geometrical representation of the system state associated with this point called a **representative point** or a **phase point**. The N-dimensional space of states is called the **phase space** of a DS. The evolution of the system state in time is portrayed by motion of the phase point along a certain curve called a **phase trajectory**. A velocity vector field is defined in the phase space of system (1.8). The vector field components are given by the right-hand sides of Eqs. (1.8):

$$[f_1(x_1, x_2, \ldots, x_N), f_2(x_1, x_2, \ldots, x_N), \ldots, f_N(x_1, x_2, \ldots, x_N)]. \tag{1.10}$$

One of the most common methods used for study of oscillatory processes is analysis of phase trajectories in DS's phase space. A complete set of different phase trajectories constitutes a **phase portrait** of the system. Research methodology by means of observation of the phase portrait evolution was first introduced and applied by L.I. Mandelstam and A.A. Andronov.

The term 'phase portrait' should be clarified on the example of a linear oscillator with dissipation:

$$\ddot{x} + 2\delta\dot{x} + x = 0. \tag{1.11}$$

First, Eq. (1.11) is transformed into form (1.8) by the introduction of a new variable $y = \dot{x} = dx/dt$. Then the second derivative becomes $\ddot{x} = d^2x/dt^2 = \dot{y}$. Then one can rewrite Eq. (1.11) as a system of two ordinary differential equations:

$$\begin{cases} \dot{x} = y, \\ \dot{y} = -2\delta y - x. \end{cases} \tag{1.12}$$

The phase space of system (1.12) is the coordinate plane (x, y). The system represents a conservative oscillator at $\delta = 0$. Oscillations of system (1.12) do not decay and their amplitude and phase are defined by initial conditions. In the context of the phase space the initial conditions are coordinates of the initial point A(x_0,y_0) on the phase plane. A trajectory in the phase space corresponding to oscillations of system (1.12) traces a circle (see Fig. 1.3a) with the center at the coordinate origin. In the center point a velocity vector is zero. Such points are called a **singular points** or an **equilibrium points**. The coordinate origin in the phase space of system (1.12) is an equilibrium point called a **center**.

At a positive value of parameter δ, system (1.12) is dissipative, and the oscillations become damped. In case $0 < \delta < 1$ phase trajectories trace twisted spirals. The phase point is attracted to the origin (Fig. 1.3b). In this case the coordinate origin is an equilibrium point called a **stable focus**. In case $\delta > 1$ the system demonstrates an aperiodic damped oscillatory process. The phase points tends to the coordinate origin along trajectories being differ from spirals (Fig. 1.3c). In this case the coordinate origin is an equilibrium point called a **stable node**. Equilibrium points stable focus and stable node are attractors. An **attractor** is a limit set

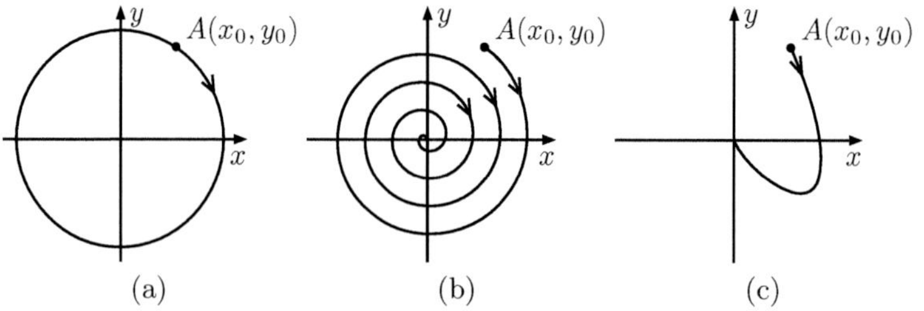

Fig. 1.3 Phase portrait of system (1.12) at $\delta = 0$ (panel (**a**)), $0 < \delta < 1$ (panel (**b**)), $\delta > 1$ (panel (**c**))

attracting all trajectories in some area of the phase space. This area is called a **basin of attraction**.

1.4 Stability of Trajectories in the Phase Space

Generally, the sense of stability is intuitively clear for everybody. In the context of DSs, stability means the property to revert to an initial oscillatory regime after removal of the force inducing the perturbation. The property of stability can be illustrated on an example of a ball moving in a well[2] (Fig. 1.4a). The well bottom at $x = x_0$ is stable, because any displacement from the point x_0 always decays over time. The opposite situation is shown in Fig. 1.4b, where the state at the peak point is unstable. One can theoretically imagine that it is possible to ideally put the ball at the highest point $x = x_0$. In practice, the ball can be put at the point x_0 with a finite accuracy. Displacements from the point x_0 inevitably present. These displacements increase over time, and the ball rolls down.

There are many different definitions of stability. The most frequently encountered ones are stability according to Poisson, stability according to Lyapunov, and asymptotic stability.

Let us consider the stability of a trajectory $\vec{x}^0(t)$ being a solution of Eq. (1.8) corresponding to certain initial conditions.

Stability according to Poisson means that the phase trajectory returns to an arbitrary small neighbourhood of the initial point $\vec{x}_0^0 = \vec{x}^0(t_0)$ after some time. If the system is reversible, the trajectory returns both forward and backward in time. The time interval after which the trajectory returns to a neighbourhood of the point $\vec{x}_0^0$ with a given radius ε is called the **Poincaré recurrence time**. An example of an open curve being stable according to Poisson is depicted in Fig. 1.5a. Stability according to Poisson is an important but weak definition. It does not contain information about the behaviour of trajectories being initially close to $\vec{x}^0(t)$. In addition, it does not involve the time-evolution of deviations from $\vec{x}^0(t)$. The behaviour of small perturbations is taken into consideration in the frameworks of Lyapunov stability formulation and the definition of asymptotic stability.

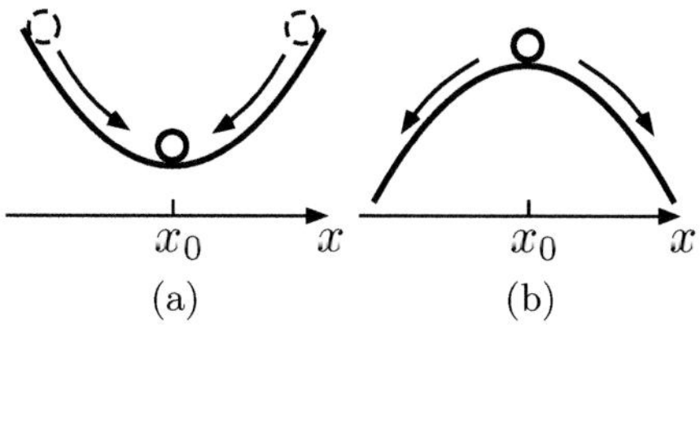

Fig. 1.4 Illustration of stable (panel (**a**)) and unstable (panel (**b**)) states

[2] This approach is a mechanical interpretation of the general problem called motion of a particle in a potential field.

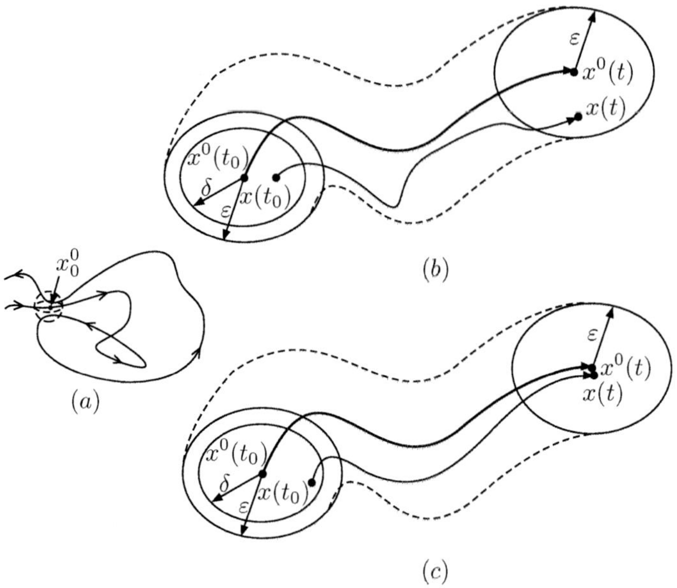

Fig. 1.5 Trajectory being stable according to Poisson (panel (**a**)), stable according to Lyapunov (panel (**b**)) and asymptotically stable (panel (**c**))

The trajectory $\vec{x}^0(t)$ is called **stable according to Lyapunov** if for any arbitrary small $\varepsilon > 0$ there exists $\delta(\varepsilon) > 0$ such that, for any trajectory $\vec{x}(t)$ for which $||\vec{x}(t_0) - \vec{x}^0(t_0)|| < \delta$, the following condition is satisfied for all values $t > t_0$: $||\vec{x}(t) - \vec{x}^0(t)|| < \varepsilon$. The symbol $||\dots||$ is the vector norm in $\mathbb{R}^N$. In other words, it means that a small initial deviation from a phase trajectory being stable according to Lyapunov does not grow over time (Fig. 1.5b). If a small perturbation δ becomes smaller and smaller over time (see Fig. 1.5c), i.e. $||\vec{x}(t) - \vec{x}^0(t)|| \to 0$ at $t \to \infty$, then the trajectory is called **asymptotically stable**. If a trajectory is asymptotically stable, it is also stable according to Lyapunov. But the opposite statement is incorrect.

The stability of a partial solution $\vec{x}^0(t)$ of Eqs. (1.8) can be explored in the frameworks of the linear stability analysis. For this purpose, one introduces a small deviation from the partial solution $\vec{y} = \vec{x}(t) - \vec{x}^0(t)$, where $\vec{x}(t)$ is a perturbed solution. After that, the linearization matrix $\hat{A}$ with elements

$$a_{j,k} = \left.\frac{df_j}{dx_k}\right|_{\vec{x}(t)=\vec{x}^0(t)}, \quad j, k = 1, 2, \dots, N \tag{1.13}$$

is introduced. Then eigenvalues of the matrix $\hat{A}$ are found by solution of the characteristic equation:

$$\mathrm{Det}\left[\hat{A}(t) - s\hat{E}\right] = 0, \tag{1.14}$$

where E is the unit matrix. The N eigenvalues correspond to N linearly independent eigenvectors $\vec{e}_i(t)$ changing their directions during the motion along the trajectory $\vec{x}^0(t)$. Using the eigenvalues of the linearization matrix, one can characterize the evolution of the perturbation $\vec{y}$ in time.

The Lyapunov characteristic exponent λ_i is a helpful tool defining the stability of a trajectory along i-th eigenvector of the matrix $\hat{A}$:

$$\lambda_i = \overline{\lim_{t\to\infty}} \frac{1}{t - t_0} \ln \left\| \frac{\vec{y}^i(t)}{\vec{y}^i(t_0)} \right\|, \tag{1.15}$$

where the bar means the upper limit, t_0 is the initial time moment. The stability of a trajectory in a space $\mathbb{R}^N$ is characterized by a set of N Lyapunov exponents. Arranged in decreasing order, the real numbers λ_i make up the Lyapunov characteristic exponent spectrum of the trajectory $\vec{x}^0(t)$. If at least one Lyapunov exponent λ_i is positive then the regime corresponding to the partial solution $\vec{x}^0(t)$ is unstable.

The stability of trajectories belonging to limit sets (for example, to attractors) is an especially important characteristic for study of DSs. Change of the limit set stability can give rise to transformation of the system dynamics. Different kinds of the limit sets are considered in the further section as well as methods of stability analysis for each of them.

1.5 Limit Sets: Classification and Stability

Study of phase portraits implies finding and analysis of phase trajectories on limit sets. The simplest examples of the limit sets are:

- **Equilibrium points (or fixed points, or equilibria, or steady states):** they correspond to quiescent states of DSs, values of dynamical variables are constant in time;
- **Limit cycles:** isolated closed curves corresponding to periodic self-oscillations;
- **Separatrix curves and surfaces:** manifolds of saddle equilibria or saddle limit cycles separating basins of attraction of stable limit sets (for instance, stable fixed points or stable limit cycles).

Issues of classification and stability analysis of the equilibrium points and limit cycles are described below. Separatrix manifolds are not individually considered in this section, but are mentioned jointly with the limit cycles and equilibrium points in the framework of description of bifurcations.

1.5.1 Equilibrium Points

Let us consider system (1.8). A point of equilibrium correspond to the absence of any motions. Then a velocity vector is zero: $\left(\frac{dx_1}{dt}, \frac{dx_2}{dt}, \ldots, \frac{dx_N}{dt}\right) = 0$. Applying this condition to the right-hand sides of Eqs. (1.8), one can find coordinates of a fixed point. Methodology of analysis of the fixed point stability is based on study of the dynamics of a linearized system in a neighbourhood of the equilibrium point. Let us assume that the point P with the coordinates $x_1 = \bar{x}_1, x_2 = \bar{x}_2, \ldots, x_N = \bar{x}_N$ is an equilibrium point. Then the right-hand sides of Eqs. (1.8) become:

$$f_i(\bar{x}_1, \bar{x}_2, \ldots, \bar{x}_N) = 0, (i = 1, \ldots, N). \tag{1.16}$$

Next, initial conditions in a neighbourhood of the point P are chosen:

$$x_1 = \bar{x}_1 + \tilde{x}_1, x_2 = \bar{x}_2 + \tilde{x}_2, \ldots, x_N = \bar{x}_N + \tilde{x}_N, \tag{1.17}$$

where $\bar{x}_1, \bar{x}_2, \ldots, \bar{x}_N$ are the coordinates of the steady state P, $\tilde{x}_1, \tilde{x}_2, \ldots, \tilde{x}_N$ are small quantities. Introducing (1.17) into (1.8), one obtains:

$$\frac{d\bar{x}_i}{dt} + \frac{d\tilde{x}_i}{dt} = f_i\left(\bar{x}_1 + \tilde{x}_1, \bar{x}_2 + \tilde{x}_2, \ldots, \bar{x}_N + \tilde{x}_N\right), \quad i = 1, \ldots, N \tag{1.18}$$

The functions $f_i(x_1, \ldots, x_N)$ can be expanded in the Taylor series close to the point $P(\bar{x}_1, \ldots, \bar{x}_N)$ in a case of small displacements $\tilde{x}_i$. Only linear terms of the expansion are used (linearization condition):

$$f_i(x_1, \ldots, x_N) = f_i\left(\bar{x}_1, \ldots, \bar{x}_N\right) + \sum_{j=1}^{N} \left.\frac{df_i(x_1, \ldots, x_N)}{dx_j}\right|_{\bar{x}_1, \ldots, \bar{x}_N} \tilde{x}_j, \tag{1.19}$$

where the derivatives are calculated at the point of equilibrium. Introduction of (1.19) into (1.18) with using condition (1.16) allows to derive equations for the evolution of perturbations:

$$\frac{d\tilde{x}_i}{dt} = \sum_{j=1}^{N} \left.\frac{df_i(x_1, \ldots, x_N)}{dx_j}\right|_{\bar{x}_1, \ldots, \bar{x}_N} \tilde{x}_j. \tag{1.20}$$

In a case of an one-dimensional DS described by the equation $\frac{dx}{dt} = f(x)$, Eq. (1.20) is transformed into:

$$\frac{d\tilde{x}}{dt} = \left.\frac{df(x)}{dx}\right|_{\bar{x}} \tilde{x}, \tag{1.21}$$

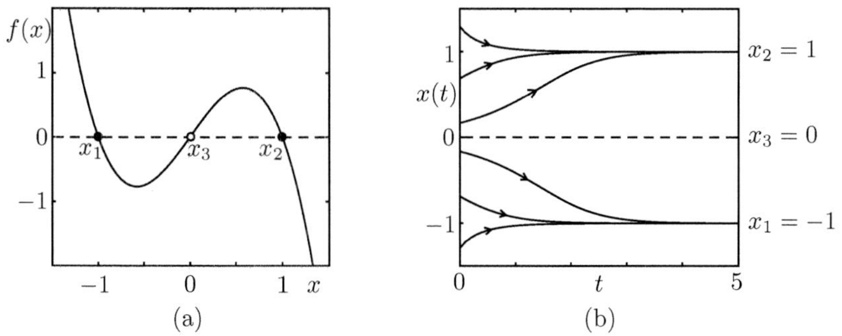

Fig. 1.6 (**a**) Function of the right-hand side of Eq. (1.22); (**b**) Solutions of Eq. (1.22) corresponding to different initial conditions

where the derivative $\frac{df(x)}{dx}$ is calculated at the point of equilibrium $\bar{x}$. The perturbation $\tilde{x}$ decreases in time for $\frac{df(x)}{dx} < 0$ and increases in case $\frac{df(x)}{dx} > 0$. Then the fixed point $\bar{x}$ is stable in case $\left.\frac{df}{dx}\right|_{\bar{x}} < 0$ and unstable in case $\left.\frac{df}{dx}\right|_{\bar{x}} > 0$.

Let us demonstrate the discussed method of steady state stability analysis on an example of an one-dimensional bistable oscillator written in the form

$$\frac{dx}{dt} = x - x^3. \tag{1.22}$$

Equilibrium points are found using the condition $\frac{dx}{dt} = 0$. Then the coordinates of the fixed points are roots of the equation $f(x) = x - x^3 = 0$. There are three roots $x_{1,2} = \pm 1$, $x_3 = 0$ (Fig. 1.6a). Values of the derivative $\frac{df(x)}{dx} = s = 1 - 3x^2$ at points $x_{1,2}$ and x_3 are $s_{1,2} = -2$ and $s_3 = 1$ correspondingly. It means the points $x_{1,2}$ are stable while the steady state x_3 is unstable. This result is confirmed by numerical simulation (see Fig. 1.6b). Depending upon the initial condition, all the solutions of Eq. (1.22) are attracted to states $x_{1,2} = \pm 1$ and repelled from the state $x_3 = 0$. Thus, oscillator (1.22) exhibits the coexistence of two attractors called **bistability**.

In a case of a multi-dimensional DS the equilibrium point P is stable if any perturbations $\tilde{x}_i$ are damped in time. It is known that solutions of linear equations (1.20) have the form $\tilde{x}_i \sim \exp(\lambda_i t)$. Then the solution $\tilde{x}_i$ is damped in two cases: all the exponent factors λ_i are real and negative or they are complex numbers with negative real parts. The factors λ_i are called **eigenvalues of the steady state** P. They are roots of the characteristic equation:

$$\det\left[\frac{df_i(x_1, \ldots, x_N)}{dx_j} - \delta_{i,j}\lambda\right] = 0, \tag{1.23}$$

where $\delta_{i,j}$ are elements of the unit matrix.

Let us consider a classification of steady states of two-dimensional DSs based on the stability analysis. The system under study is considered in the form:

$$\begin{cases} \dfrac{dx}{dt} = f(x, y), \\ \\ \dfrac{dy}{dt} = g(x, y). \end{cases} \tag{1.24}$$

The point P with coordinates $(\bar{x},\bar{y})$ is assumed to be an equilibrium point. Equations describing the evolution of perturbations $\tilde{x}$ and $\tilde{y}$ in a neighbourhood of the point P (two-dimensional form of Eq. (1.20)) become:

$$\begin{cases} \dfrac{d\tilde{x}}{dt} = A\tilde{x} + B\tilde{y}, \\ \\ \dfrac{d\tilde{y}}{dt} = C\tilde{x} + D\tilde{y}, \end{cases} \tag{1.25}$$

or in the matrix form:

$$\begin{bmatrix} \dfrac{d\tilde{x}}{dt} \\ \\ \dfrac{d\tilde{y}}{dt} \end{bmatrix} = \begin{bmatrix} A & B \\ \\ C & D \end{bmatrix} \begin{bmatrix} \tilde{x} \\ \\ \tilde{y} \end{bmatrix}, \tag{1.26}$$

where $A = \frac{df(x,y)}{dx}\Big|_{\bar{x},\bar{y}}$, $B = \frac{df(x,y)}{dy}\Big|_{\bar{x},\bar{y}}$, $C = \frac{dg(x,y)}{dx}\Big|_{\bar{x},\bar{y}}$, $D = \frac{dg(x,y)}{dy}\Big|_{\bar{x},\bar{y}}$. The corresponding characteristic equation (1.23) becomes:

$$\det \begin{bmatrix} A - \lambda & B \\ \\ C & D - \lambda \end{bmatrix} = 0. \tag{1.27}$$

One can rewrite Eq. (1.27) as $\lambda^2 - K_1\lambda + K_2 = 0$, where $K_1 = A + D$ and $K_2 = AD - BC$. Then roots of the characteristic equation are:

$$\lambda_{1,2} = \frac{K_1}{2} \pm \sqrt{\frac{K_1^2}{4} - K_2}. \tag{1.28}$$

Depending on the parameters K_1 and K_2, the eigenvalues $\lambda_{1,2}$ possess different complex or real values. It determines classification of equilibrium points. The classification of equilibria of two-dimensional DSs is presented below and illustrated in Fig. 1.7:

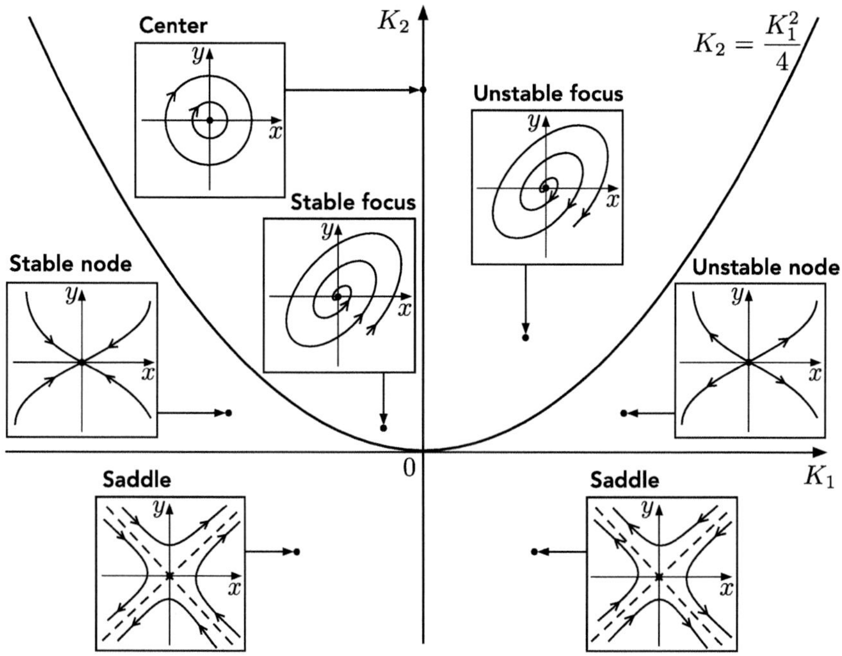

Fig. 1.7 Different kinds of steady states on the plane (K_1, K_2). Insets show the behaviour of phase trajectories in the neighbourhood of each equilibrium

- **Stable node** is observed in a case of real negative values $\lambda_{1,2}$. Phase trajectories are attracted to the fixed point and describe aperiodic process. Deviations from the point of equilibrium are fast-decreasing in time.
- **Unstable node** is observed in a case of real positive values $\lambda_{1,2}$. Phase trajectories are repelled from the fixed point and describe aperiodic process. Deviations from the point of equilibrium are fast-increasing in time.
- **Stable focus** corresponds to complex eigenvalues with the negative real part. Phase trajectories in a neighbourhood of the steady state trace twisted spirals. Time realizations of dynamical variables describe a damped oscillatory process. The oscillation amplitude decreases in time;
- **Unstable focus** corresponds to complex eigenvalues with a positive real part. Phase trajectories in the neighbourhood of the steady state trace untwisted spirals. Oscillations of dynamical variables describe a periodic process with the amplitude increasing in time.
- **Saddle** has two real eigenvalues: positive and negative. The saddle point of equilibrium has stable and unstable manifolds (dashed lines in Fig. 1.7). A phase trajectory at the stable manifold is attracted to the saddle, while a phase trajectory at the unstable manifold is repelled from the equilibrium. Phase trajectories starting close to the stable manifold are initially attracted to the fixed point during finite time, but after they move away from the steady state.

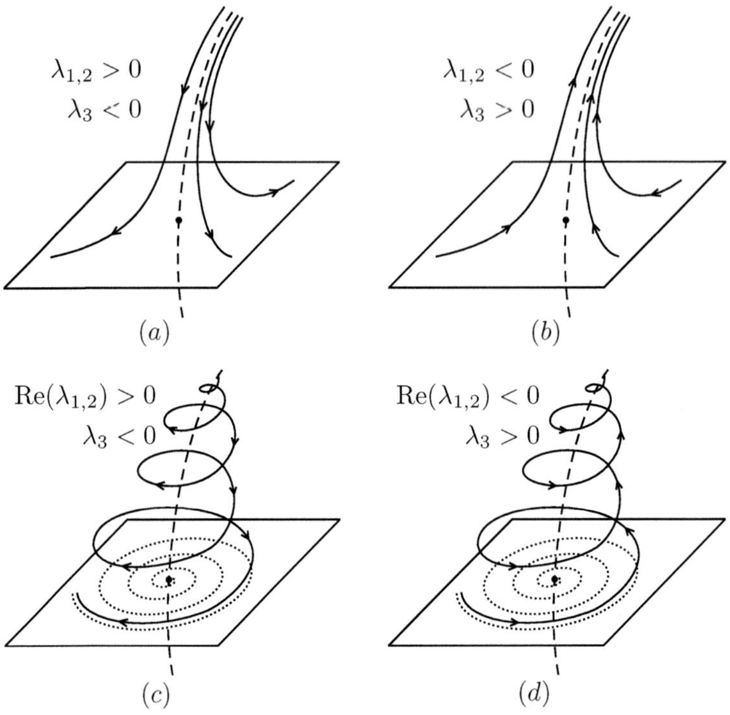

Fig. 1.8 Phase trajectories in a neighbourhood of a saddle-node (panels (**a**) and (**b**)) and a saddle-focus (panels (**c**) and (**d**)) for different sets of eigenvalues. Dashed lines indicate stable (panels (**a**) and (**c**)) and unstable (panels (**b**) and (**d**)) manifolds of the saddle point

- **Center** is observed in the phase space of conservative oscillators. Eigenvalues are pure imaginary. Phase trajectories are closed curves and describe non-damped periodic process. The amplitude of oscillations is constant and defined by initial conditions.

In a case of three-dimensional DSs Eq. (1.23) has three roots, and points of equilibrium are characterized by three eigenvalues $\lambda_{1,2,3}$. It extends a manifold of saddle-type equilibria. The following configurations are possible in the three-dimensional phase space besides centres, focuses, and nodes:

- **Saddle-node** corresponds to three real eigenvalues of different sign. Phase trajectories behave similarly to the case of a saddle in the 2D phase space (Fig. 1.8a, b). Initially attracted to the steady state, they change their direction after finite time and move away from the equilibria. Positive eigenvalues $\lambda_i > 0$ determine the dimension of an unstable manifold, while negative eigenvalues $\lambda_i < 0$ determine the dimension of the stable one.
- **Saddle-focus** has one real and two complex-conjugate eigenvalues. Signs of the real eigenvalue and a real part of the complex-conjugate eigenvalues are opposite.

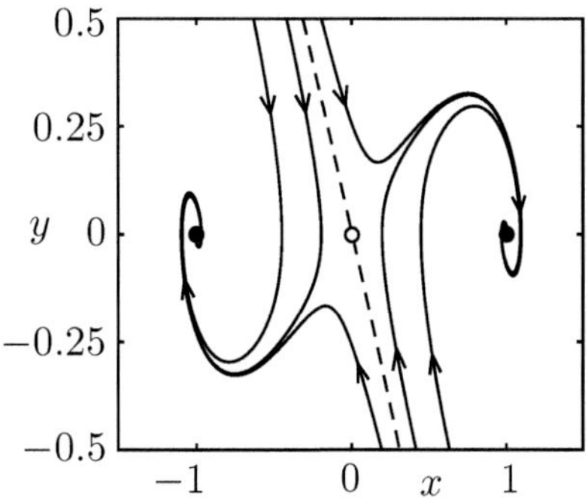

Fig. 1.9 Phase space of system (1.29). The filled circles are stable focuses and the empty circle at the origin is a saddle. The dashed line is a separatrix of the saddle

One can distinguish a plane in the phase space, where projections of the phase trajectories trace twisted or untwisted spirals (Fig. 1.8c, d).

Let us demonstrate the steady state stability analysis in a case of two-dimensional DSs on a example of the bistable oscillator with finite dissipation given by the equation $\frac{d^2x}{dt^2} + \frac{dx}{dt} - x + x^3 = 0$. This equation is rewritten as a system of two first-order differential equations:

$$\begin{cases} \dfrac{dx}{dt} = y, \\ \\ \dfrac{dy}{dt} = -y + x - x^3. \end{cases} \tag{1.29}$$

Equilibrium points are found using the conditions $\frac{dx}{dt} = 0$ and $\frac{dy}{dt} = 0$. Thus, fixed points on the phase plane (x,y) are $P_1(-1, 0)$, $P_2(1, 0)$, $P_3(0, 0)$ (see Fig. 1.9). Elements A-D of characteristic equation (1.27) are: $A = 0$, $B = 1$, $C = 1 - 3x^2$, $D = -1$. The factor C possesses two values: $C_{1,2} = -2$ for the equilibria $P_{1,2}$ and $C_3 = 1$ for the steady state P_3. The characteristic equations are $\lambda^2 + \lambda + 2 = 0$ for the points $P_{1,2}$ and $\lambda^2 + \lambda - 1 = 0$ for the point P_3. Then the eigenvalues of the points $P_{1,2}$ are $\lambda_{1,2} = -0.5 \pm i\sqrt{1.75}$, while the eigenvalues for the point P_3 are $\lambda_{1,2} = -0.5 \pm \sqrt{1.25}$. It means the points $P_{1,2}$ are stable focuses and P_3 is a saddle point. This result is confirmed by numerical modelling system (1.29). All the trajectories are attracted to either stable equilibria P_1 or P_2 and trace spirals in a neighbourhood of the attractors (Fig. 1.9). The behaviour of trajectories close to the equilibrium point P_3 is typical for a saddle. Initially, the trajectories tend to the saddle, but after some finite time they change their direction to the attractors. The dashed line in Fig. 1.9 is a stable manifold of the saddle. In addition, it is a boundary between basins of attraction of the points P_1 and P_2. This line is called a **separatrix of a saddle**.

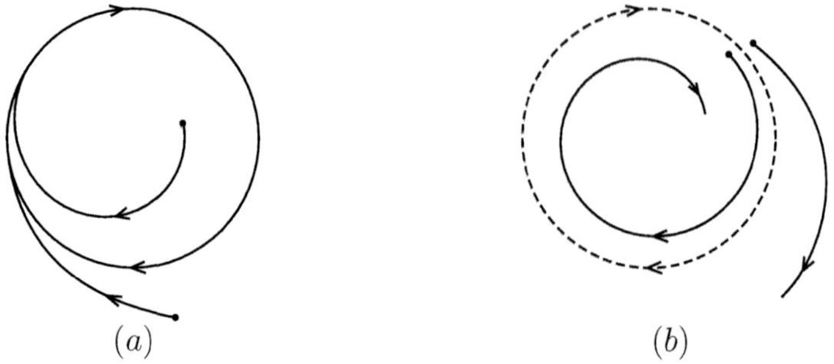

Fig. 1.10 Stable (panel (**a**)) and unstable (panel (**b**)) limit cycles on the phase plane and associated trajectories

1.5.2 Limit Cycles

Periodic solutions of system (1.8) satisfy the condition $\vec{x}^0(t) = \vec{x}^0(t+T)$, where T is the period of the solution. A periodic solution is isolated if it has a finite-size vicinity not including any other solutions. An isolated periodic solutions of an autonomous DS is illustrated by a **limit cycle** in the phase space. The limit cycle is an isolated closed curve having no self-intersections. The limit cycles can be observed in the phase space of DSs with dimension $N \geq 2$. Classification of limit cycles is based on stability analysis. Thus, two kinds of limit cycles can be observed in the phase space of two-dimensional DSs: stable limit cycles and unstable ones. The stable limit cycle is an image of regular self-oscillations. The self-oscillatory regime is stationary and stable, and the corresponding limit cycle is an attractor (Fig. 1.10a). The unstable limit cycle on a phase plane is a repeller (Fig. 1.10b) and usually separates basins of attraction of other attractors imaging self-oscillatory or quiescent steady-state regimes.

One can characterize the stability of a periodic solution by exploration of the evolution of small perturbations in the framework of the linear stability analysis described in Sect. 1.4. The linearization matrix $\hat{A}$ calculated at points of the trajectory corresponding to the periodic solution $\vec{x}^0(t)$ is also periodic: $\hat{A}(t) = \hat{A}(t+T)$. It means that an equation for the perturbation evolution (it is Eq. (1.20), but applied to a case of a periodic solution) is linear with periodic coefficients. If the evolution of small perturbations $\vec{y}$ over the period T is known, the solution stability can be revealed. The perturbation evolution is represented as being:

$$\vec{y}(t_0 + T) = \hat{M}_T \vec{y}(t_0), \tag{1.30}$$

where $\hat{M}_T$ is independent of time and is called a **monodromy matrix**. The eigenvalues of the monodromy matrix are called **multipliers of a periodic solution**. They are roots of the characteristic equations:

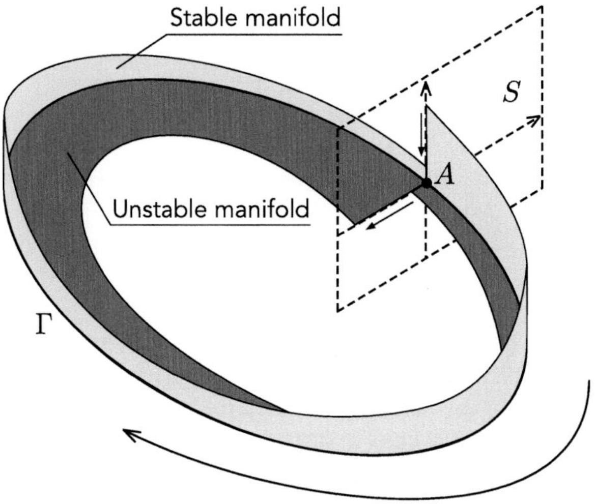

Fig. 1.11 Illustration to stability analysis of a limit cycle in the three-dimensional phase space. The plane S defines the Poincaré section. The point A is a point of intersection of the plane S and the limit cycle Γ

$$\mathrm{Det}\left[\hat{M}_T - \mu\hat{E}\right] = 0. \tag{1.31}$$

The action of the monodromy matrix is described by multipliers as follows. The perturbation is considered as a set of projections onto the eigenvectors of the matrix $\hat{M}_T$. Each component of the initial perturbation is multiplied by the corresponding multiplier μ_i over the period T. Then the condition for the stability of a periodic solution is that all the absolute values of multipliers except one of them, (which is always unit) are less than unity: $|\mu_i| \leq 1$. At least, one multiplier of a periodic solution always equals to unity. It is caused by exact periodicity of the solution over the period T. The multipliers are related to the Lyapunov exponents of the periodic solution: $\lambda_i = \frac{1}{T}\ln|\mu_i|$.

In a case of a two-dimensional DS, a periodic solution is characterized by two multipliers. One of them is unit. The second multiplier is greater or less than unity. For this reason limit cycles on a phase plane can be either stable or unstable. A manifold of limit cycle types in a case of multi-dimensional DSs is broader. Thus, saddle limit cycles can exist in the phase space of a three-dimensional DS.

Let us consider geometric interpretation of the monodromy operator action and a classification of limit cycles in a case of the three-dimensional phase space (Fig. 1.11). The point A on the limit cycle Γ is chosen for description. The point A belongs to the plane S being transversal to the curve Γ. Phase trajectories in a vicinity of the limit cycle Γ intersect the plane S in some point and then do it again in other point after one revolution along the curve Γ. Then we deal with an iterated map: each previous intersection corresponding to a perturbation $\vec{y}(n)$

dictates the further displacement after one revolution $\vec{y}(n+1)$. This process is determined by some evolution operator. Such a map is called the **Poincaré return map** or the **Poincaré map**. The point A at the periodic solution is an equilibrium point of the iterated map in the plane S, because it is iterated in itself. It means the stability of the periodic solution Γ is associated with the stability of the steady state A. The monodromy operator describes the evolution of small perturbations in a neighbourhood of the steady state A. Consequently, the multipliers characterize the stability of the steady state A and hence the stability of the limit cycle Γ. A manifold of repelled phase trajectories (unstable manifold in Fig. 1.11) is associated with multipliers the absolute value of which is greater than unity. A manifold of attracted phase trajectories (stable manifold in Fig. 1.11) is associated with multipliers the absolute value of which is less than unity. Thus, in the three-dimensional phase space one can distinguish stable limit cycles (all absolute values of multipliers, except of one, are less than unity), unstable limit cycles (all absolute values of multipliers, except of one, are greater than unity) and saddle limit cycles (some absolute values of multipliers are less than unity, while other ones are greater than unity, one multiplier equals to unity).

Elements of the monodromy matrix for a periodic solution of Eq. (1.8) can be estimated in numerical experiments. The solution period is known (some quantity T). One takes into consideration a point $A(\bar{x}_1, \bar{x}_2, \ldots, \bar{x}_N)$ on a limit cycle. Next, a vector of deviations from the point A is introduced in the form: $\vec{y} = [0, 0, \ldots, \alpha_i, \ldots, 0, 0]$, where a small nonzero quantity α stands at i-th position. Numerical modelling of system (1.8) is carried out from the initial conditions $A^*(\bar{x}_1, \bar{x}_2, \ldots, \bar{x}_i + \alpha_i, \ldots, \bar{x}_N)$ corresponding to a perturbed state. System (1.8) is simulated in the time interval of one period T. Then the new vector of deviations from the point A after one period T becomes equal to the i-th column of the monodromy matrix multiplied by α_i. Indeed, rewriting equation for perturbations (1.30) for case $\vec{y}(t_0) = [0, 0, \ldots, \alpha_i, \ldots, 0, 0]$, one obtains:

$$\vec{y}(t_0+T) = \begin{bmatrix} y_1(t_0+T) \\ \vdots \\ \vdots \\ \vdots \\ y_N(t_0+T) \end{bmatrix} = \begin{bmatrix} M_{1,1} & \cdots\cdots & M_{1,N} \\ \vdots & & \vdots \\ \vdots & & \vdots \\ \vdots & & \vdots \\ M_{N,1} & \cdots\cdots & M_{N,N} \end{bmatrix} \begin{bmatrix} 0 \\ \vdots \\ \alpha_i \\ \vdots \\ 0 \end{bmatrix}$$

$$= \begin{bmatrix} \alpha_i M_{1,i} \\ \vdots \\ \vdots \\ \vdots \\ \alpha_i M_{N,i} \end{bmatrix} \tag{1.32}$$

Changing the position of the quantity α_i, all the elements of the monodromy matrix can be calculated.

1.6 Bifurcations

Generally, the term 'bifurcation' means qualitative changes in the behaviour of DSs caused by variation of parameters. In terms of qualitative theory of differential equations, the bifurcations are associated with transformations of limit sets in the phase space: appearance, disappearance, change of the stability character. Such transformations can be hidden for experimenters in some cases. The bifurcations can involve limit sets which are not attractors and therefore cannot be experimentally observed. In addition, the bifurcations can occur out of a vicinity of a regime observed by the experimenter at the bifurcation moment in a case of the coexistence of several attractors.

Bifurcations in deterministic DSs are divided into local bifurcations and nonlocal ones. The local bifurcations are associated with the behaviour of phase trajectories in a local neighbourhood of limit sets. The local bifurcations are related to transformations of certain limit sets and can be identified by means of linear stability analysis. Nonlocal bifurcations involve manifolds of saddle limit sets (separatrix loops, separatrix curves and surfaces, homoclinic and heteroclinic curves). Usually, linear approach is insufficient for description of these bifurcations. In such a case nonlinear properties cannot be neglected and must be included into consideration.

In a general case, the occurrence of bifurcations observed in multi-dimensional DSs can require satisfaction of several conditions imposed on parameters of the system. Then the characteristic of a bifurcation called the **codimension** is equal to the number of such conditions.

DSs are not robust at the bifurcation point: a small change of a parameter in any direction leads to principal changes in the dynamics. One can consider the DS exactly at the moment of a bifurcation only in the framework of theoretical consideration. In contrast, the parameters of DSs cannot be fixed at the bifurcation values with absolute accuracy in real experiments. An experimenter can detect the bifurcation values of the parameters only with finite accuracy.

Bifurcations in deterministic DS are classified as soft or hard (crises). The soft bifurcations are characterized by continuous changes of the dynamics and occur in the same manner in a case of increasing and decreasing the main parameter. The hard bifurcations induce sharp changes in the dynamics. They are characterised by hysteresis and occur in different manner depending on the direction of the main parameter change.

In contrast to bifurcations in deterministic DSs, stochastic bifurcations (i.e. bifurcations observed in the presence of noise) do not possess the property of universality. In each particular case, features of the stochastic bifurcations are dictated by intrinsic peculiarities of the system under study. Due to this fact, a complete general classification of stochastic bifurcations is absent. Nevertheless, depending on a consideration approach, one can distinguish two kinds of stochastic bifurcations [6]: P-bifurcations and D-bifurcations. The P-bifurcations are associated with qualitative changes of the probability density function of dynamical variables. The D-bifurcations are associated with the evolution of the Lyapunov exponents when the noise intensity or system parameters are varied.

The most common bifurcations of steady states and limit cycles observed in deterministic DSs are described below in the further subsections. The bifurcations of equilibria are considered on the examples of normal forms, i.e. the simplest DS exhibiting the bifurcation.

1.6.1 Bifurcations of Equilibrium Points

Saddle-Node Bifurcation

Let us assume that there are two equilibrium points (a saddle and a stable node) in the phase space of a two-dimensional DS (Fig. 1.12a). Changing a system parameter m leads to convergence of the steady states and merging into one point called a saddle-node at a bifurcation point $m = m_*$ (Fig. 1.12b). The steady states disappear after passing through the bifurcation value m_* (Fig. 1.12c).

The simplest DS for the saddle-node bifurcation of steady states is a model

$$\frac{dx}{dt} = m - x^2. \tag{1.33}$$

System (1.33) is one-dimensional. Technically, it cannot realize the saddle-node bifurcation (it requires at least the two-dimensional phase space). However, bifurcation changes in Eq. (1.33) caused by varying a parameter m fully replicate the character of the saddle-node bifurcation. Equilibria of Eq. (1.33) are $x_1 = \sqrt{m}$ and $x_2 = -\sqrt{m}$. It is evident that the points $x_{1,2}$ exist only in case $m > 0$ (coordinates of steady states are always real quantities). The stability of the fixed points $x_{1,2}$ is determined by a derivative of the right-hand side of Eq. (1.33) $\frac{df}{dx} = -2x$ calculated

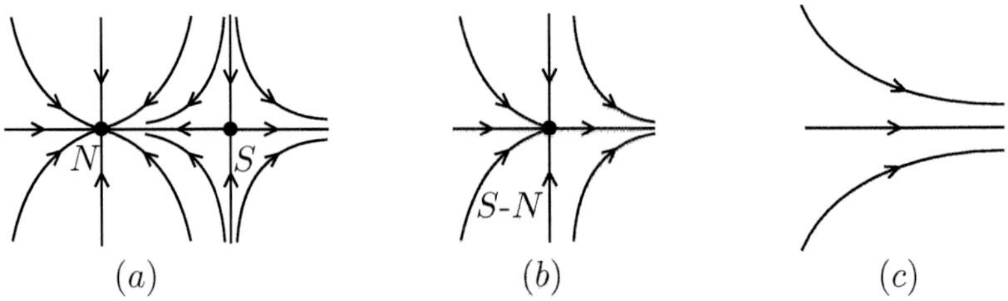

Fig. 1.12 Evolution of the phase space when a saddle-node bifurcation occurs: before the bifurcation (panel (**a**)), at the bifurcation moment (panel (**b**)) and after the bifurcation (panel (**c**))

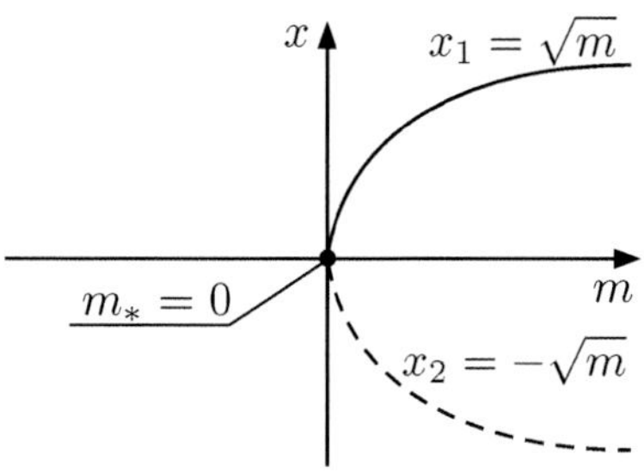

Fig. 1.13 Bifurcation diagram for system (1.33). The solid line marks the stable steady state x_1, the dashed line marks the unstable equilibrium x_2

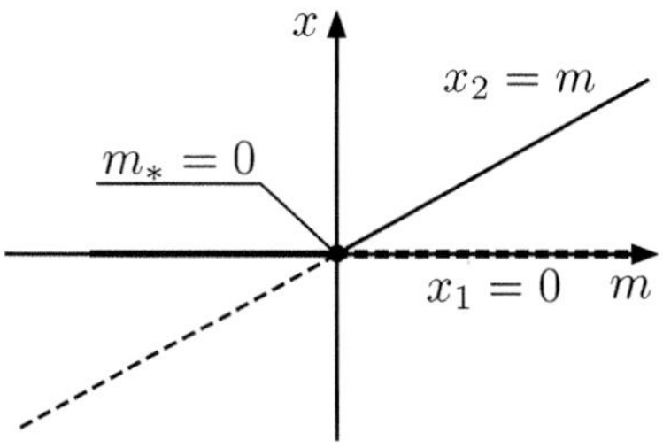

Fig. 1.14 Bifurcation diagram for system (1.34). The solid line marks the stable steady state, the dashed line marks the unstable one

at the points $x_{1,2}$. At the equilibrium x_1 the derivative becomes $\frac{df}{dx} = -2\sqrt{m}$ and is always negative when the steady state x_1 exists. It means the steady state x_1 is stable for any $m > 0$. Similarly, a value of the derivative at the equilibrium x_2 is $\frac{df}{dx} = 2\sqrt{m}$. It is always positive when the steady state x_2 exists. Then, the steady state x_2 is unstable for any $m > 0$. In summary, the following bifurcation diagram is obtained (Fig. 1.13). In case $m > 0$ there exist two equilibria (stable and unstable) in the phase space. Decreasing the parameter m gives rise to convergence of the fixed points and merging at the bifurcation value $m_* = 0$. After the bifurcation there is no steady states in the phase space.

Transcritical Bifurcation

A transcritical bifurcation corresponds to exchange of stability between two points of equilibrium. The normal form for this bifurcation is the equation

$$\frac{dx}{dt} = mx - x^2. \tag{1.34}$$

Steady states of system (1.34) are $x_1 = 0$ and $x_2 = m$. They exist at any values of a parameter m. A value of the derivative $\frac{df}{dx}$ defining the stability becomes $\frac{df}{dx} = m$ for the steady state x_1 and $\frac{df}{dx} = -m$ for the steady state x_2. It means the fixed point $x_1 = 0$ is stable for $m < 0$ and unstable for $m > 0$. At the same time the steady state $x_2 = m$ is stable for $m > 0$ and unstable for $m < 0$. In summary, two equilibria bifurcate at $m_* = 0$ and exchange of stability occurs (Fig. 1.14).

Fig. 1.15 Bifurcation diagram for system (1.35). The solid line marks the stable steady state, the dashed 74le one

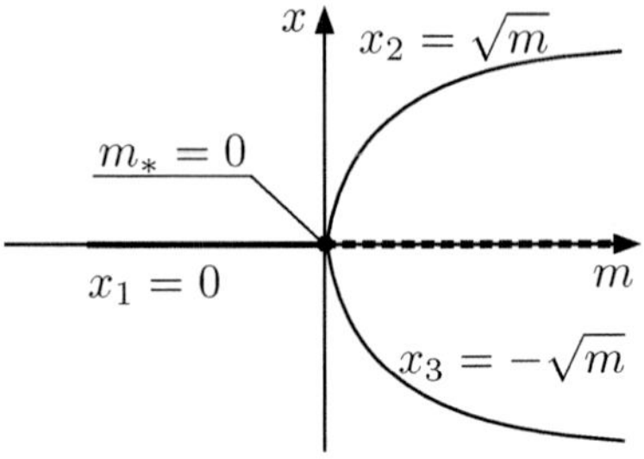

Pitchfork Bifurcation

A pitchfork bifurcation typically realizes transition from monostability to bistability. It consists in the appearance of two stable states from the point of equilibrium, which was initially stable before the bifurcation and unstable afterwards (it becomes a fully unstable steady state in a case of an one-dimensional DS and a saddle equilibrium in a two- or multi-dimensional system). The normal form for this bifurcation is the following equation:

$$\frac{dx}{dt} = mx - x^3. \tag{1.35}$$

The condition for steady state $\frac{dx}{dt} = 0$ leads to the cubic equation $mx - x^3 = 0$ which has one solution $x_1 = 0$ for $m < 0$ and three solutions $x_1 = 0$ and $x_{2,3} = \pm\sqrt{m}$ for $m > 0$. It means before the bifurcation only one steady state x_1 exists in the phase space. After the bifurcation at $m_* = 0$ two new steady states $x_{2,3}$ appear in a vicinity of the equilibrium point x_1. A value of the derivative $\frac{df}{dx}$ becomes $\frac{df}{dx} = m$ for the steady state x_1 and $\frac{df}{dx} = -2m$ for the steady states $x_{2,3}$. It means the fixed point $x_1 = 0$ is stable for $m < 0$ and unstable for $m > 0$. At the same time, the fixed points $x_{2,3}$ are stable over the whole area of existence (any positive values of the parameter m). Then one can plot a complete bifurcation diagram (Fig. 1.15). Growth of the parameter m causes the pitchfork bifurcation at $m_* = 0$. Two stable steady states $x_{2,3}$ appear in a vicinity of the steady state x_1 which loses its stability after the bifurcation.

Imperfect Pitchfork and Transcritical Bifurcations

Models (1.34) and (1.35) are absolutely symmetric. However, one cannot experimentally realize symmetric DS with absolute accuracy. Mismatches and imperfections always persist in real systems. If the difference between a numerical model and a real DS cannot be neglected, the model equations must be corrected. For example, asymmetry can be taken into consideration by addition of a constant ξ into Eqs. (1.34) and (1.35) as being:

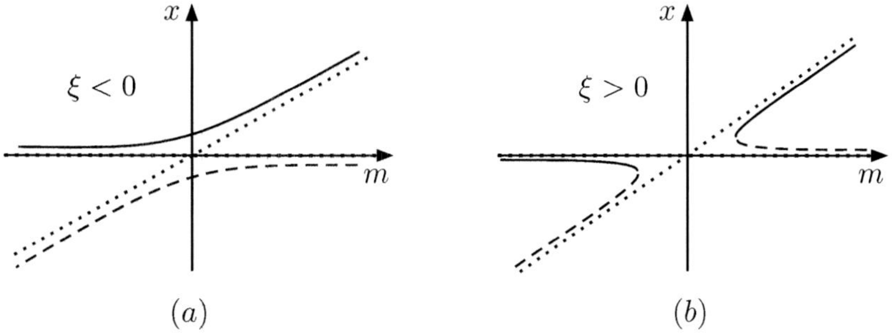

Fig. 1.16 Imperfect transcritical bifurcation in system (1.36) at $\xi < 0$ (panel (**a**)) and $\xi > 0$ (panel (**b**))

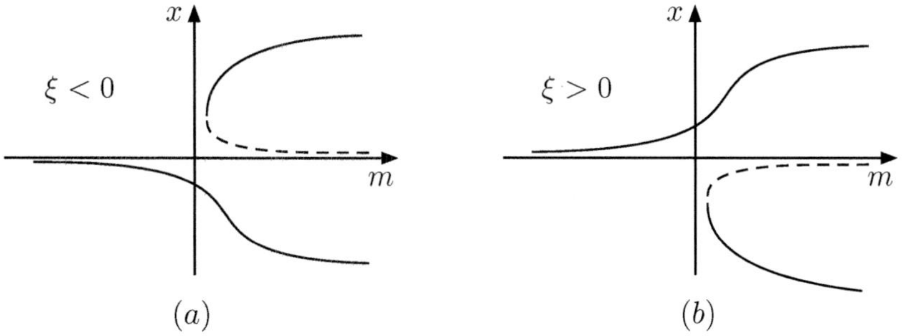

Fig. 1.17 Imperfect pitchfork bifurcation in system (1.37) at $\xi < 0$ (panel (**a**)) and $\xi > 0$ (panel (**b**))

$$\frac{dx}{dt} = mx - x^2 + \xi, \tag{1.36}$$

$$\frac{dx}{dt} = mx - x^3 + \xi. \tag{1.37}$$

Modified system (1.36) has two equilibria $x_{1,2} = -\frac{m}{2} \pm \sqrt{\frac{m^2}{4} - \xi}$, which tend to $x_1 = m$ and $x_2 = 0$ in limit case $\xi \to 0$. Depending on the sign of the quantity ξ, a bifurcation diagram for Eq. (1.36) is transformed from the ideal case depicted in Fig. 1.14 into either imperfect case in Fig. 1.16a or b. Strictly speaking, there is no bifurcation in Fig. 1.16a, and the bifurcation diagram in Fig. 1.16b contains two saddle-node bifurcations instead of one transcritical bifurcation. Nevertheless, the bifurcation diagrams in Fig. 1.16 are transformed into Fig. 1.14 in case $\xi \to 0$. For this reason, the mentioned bifurcation phenomenon is called an **imperfect transcritical bifurcation**.

Similarly, the impact of asymmetry in system (1.37) is reflected in transformation of a pitchfork bifurcation into a saddle-node bifurcation (Fig. 1.17). It results in the appearance of the second attractor and an unstable point of equilibrium. Meanwhile, the first stable equilibrium, existing before bifurcation, only changes its position without changing the stability character.

1.6.2 *Bifurcations of Limit Cycles*

A closed phase trajectory corresponding to a periodic solution is called **nondegenerate** if all its multipliers do not equal to unity (except of one multiplier being always equal to unity, which is eliminated from the further consideration). Such trajectories are robust. They do not disappear with weak change of a parameter, but become slightly distorted. In this case the behaviour of phase trajectories in a neighbourhood of the periodic solution is not principally transformed. Qualitative changes occur when one multiplier (or two complex-conjugate multipliers) passes out of the unit circle. The condition for observation of a bifurcation is $|\mu_i| = 1$. The possible options are considered below.

Andronov-Hopf Bifurcation

This bifurcation was first explored by A.A. Andronov for two-dimensional DSs and then generalized by E. Hopf to multi-dimensional systems. An Andronov-Hopf bifurcation describes excitation of self-oscillations. Hard and soft oscillation excitations are distinguished. The soft oscillation excitation is characterised by continuous change of the self-oscillation amplitude caused by varying a parameter and by the absence of hysteresis. The hard oscillation excitation is characterised by sharp increase of the self-oscillation amplitude caused by varying a parameter and the presence of hysteresis. Schematic description of the Andronov-Hopf bifurcation is presented below. A detailed analysis of the hard and soft oscillation excitation is carried out on a particular example of the Van der Pol self-oscillator in Sect. 5.2.1.

Soft self-oscillation excitation is associated with the supercritical Andronov-Hopf bifurcation. In the phase space the supercritical Andronov-Hopf bifurcation consists in the birth of a stable limit cycle from an equilibrium lost its stability (Fig. 1.18). Before the supercritical Andronov-Hopf bifurcation an attractor, a stable fixed point (stable focus), corresponding to the absence of oscillations, exists in the phase space. After the bifurcation other attractor, a stable limit cycle, appears. Changing the parameter in the opposite direction leads to continuous contraction of the stable limit cycle. Finally, the limit cycle collides into the unstable equilibrium and disappears. At this moment the unstable fixed point becomes stable.

Hard self-oscillation excitation is associated with the subcritical Andronov-Hopf bifurcation. In the phase space, the subcritical Andronov-Hopf bifurcation consists in the birth of an unstable limit cycle from an unstable equilibrium (Fig. 1.19). After the bifurcation the unstable fixed point becomes stable. Changing the parameter in the opposite direction leads to continuous contraction of the unstable limit cycle.

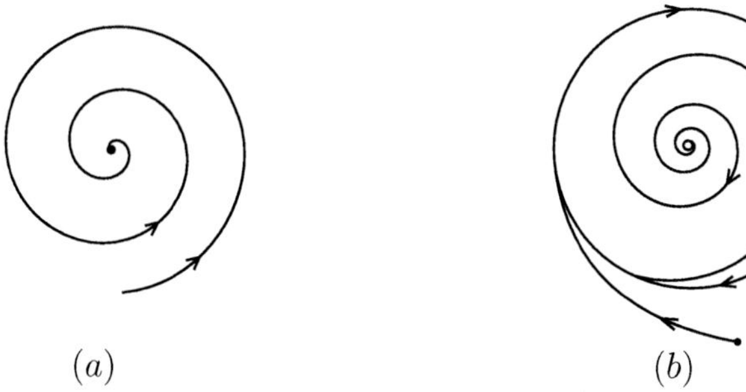

Fig. 1.18 Supercritical Andronov-Hopf bifurcation: the phase space before (panel (**a**)) and after (panel (**b**)) the bifurcation

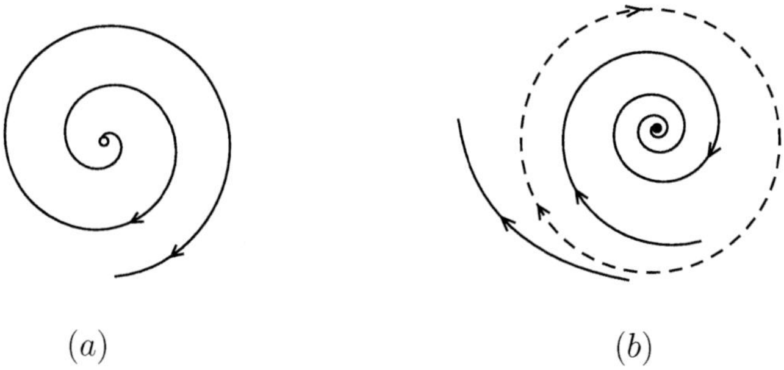

Fig. 1.19 Subcritical Andronov-Hopf bifurcation: the phase space before (panel (**a**)) and after (panel (**b**)) the bifurcation

Finally, the limit cycle collides into the stable equilibrium and disappears. At this moment, the stable fixed point becomes unstable.

Saddle-Node Bifurcation

A saddle-node bifurcation of limit cycles is illustrated in Fig. 1.20. Before the bifurcation there exist two limit cycles in the phase space: stable Γ_N and saddle Γ_S (Fig. 1.20a). In the Poincaré section two limit cycles Γ_N and Γ_S are imaged as two steady states A_N (stable node) and A_S (saddle steady state) of a map describing the evolution of perturbations. Parameter variation moving a DS to a bifurcation point gives rise to convergence of the limit cycles (and the fixed points in the Poincaré section) and merging into degenerate closed curve Γ_{SN} (Fig. 1.20b). At this moment one of multipliers of the curve Γ_{SN} becomes equal to unity (besides the multiplier which is always unit). After the bifurcation the closed curve Γ_{SN} disappears. Finally, there is no limit cycles Γ_N and Γ_S and no degenerate curve Γ_{SN} in the phase space. The described phenomenon is reflected in a saddle-node bifurcation of the points of equilibrium A_N and A_S in the Poincaré section.

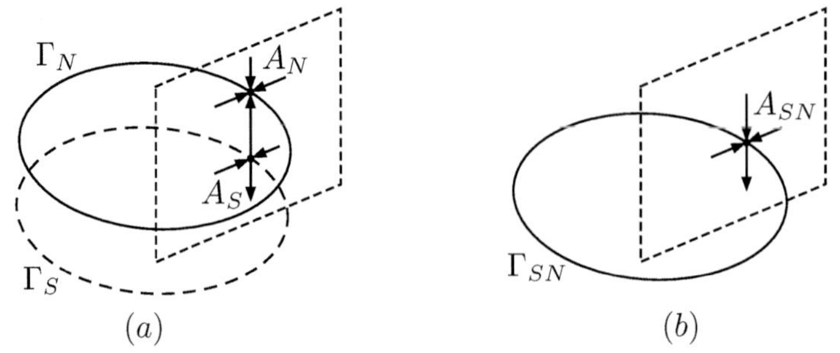

Fig. 1.20 Saddle-node bifurcation of limit cycles: the phase space before the bifurcation (panel (**a**)) and at the bifurcation moment (panel (**b**))

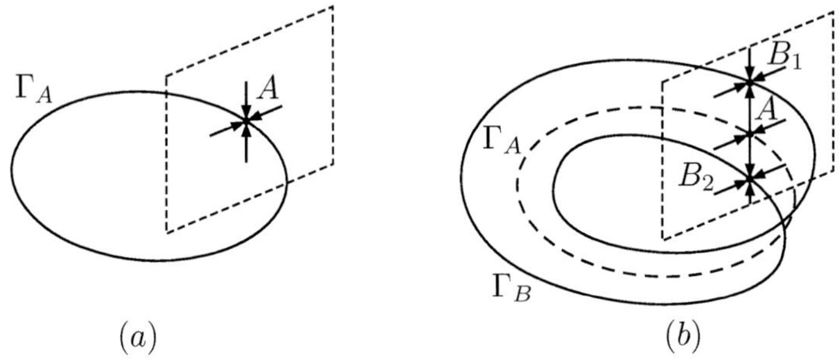

Fig. 1.21 Period-doubling bifurcation of a limit cycle: the stable limit cycle Γ_A before the bifurcation (panel (**a**)) and two limit cycles (saddle limit cycle Γ_A and stable limit cycle Γ_B) after the bifurcation (panel (**b**))

Period-Doubling Bifurcation

If one of multipliers of a stable limit cycle corresponding to self-oscillations with the period T becomes equal to minus unity, $\mu_i = -1$, then a period-doubling bifurcation occurs. In such a case a small initial perturbation y_i in the direction of the eigenvector $\vec{e}_i$ associated with the multiplier μ_i becomes $-y_i$ after one period T and again y_i after two periods. Then the perturbed phase trajectory returned into the initial point after two revolutions. After the bifurcation the limit cycle becomes unstable, but phase trajectories remain to be attracted to the neighbourhood of the limit cycle with inversion of deviations in the direction of the eigenvector $\vec{e}_i$ every period T. It becomes possible only when a new stable limit cycle with period $2T$ appears in a vicinity of the became unstable limit cycle.

The described situation is illustrated in Fig. 1.21. At the bifurcation moment, the limit cycle Γ_A being initially stable becomes saddle and in its vicinity the new stable limit cycle Γ_B appears. The limit cycle Γ_B corresponds to self-oscillations characterized by the period $2T$ and passes through the Poincaré section in two points B_1 and B_2.

The period-doubling bifurcation plays an important role in one of scenarios of transition from the regular to chaotic dynamics called a **cascade of period-doubling bifurcations**. It consists in multiple bifurcations when a stable limit cycle with the period nT becomes saddle and in its vicinity a new stable limit cycle with the period $2nT$ appears. By this way one realize self-oscillatory regimes of the periods T, $2T$, $4T$, $8T$, $16T$ ... After passing through a critical parameter value the period becomes infinite and a chaotic attractor appears. A more detailed description of such a transition to chaos is presented in the fifth chapter (Sect. 5.4).

Two-Dimensional Torus Birth

The supercritical Andronov-Hopf bifurcation resulting in the appearance of a stable limit cycle has an analog in the Poincaré section. This configuration is depicted in Fig. 1.22. The equilibrium point A imaging the stable limit cycle Γ_A (Fig. 1.22a) is characterized by two complex conjugate multipliers $\mu_1 = |\mu_1| \exp(i\omega)$, $\mu_2 = |\mu_2| \exp(-i\omega)$ with the additional condition $\omega \neq 0$, $p\pi/q$, where p and q are any positive integers (this condition excludes resonance phenomena). At the bifurcation moment, the absolute value of two complex conjugate multipliers equals to unity. After the bifurcation the steady state A becomes unstable and in its vicinity

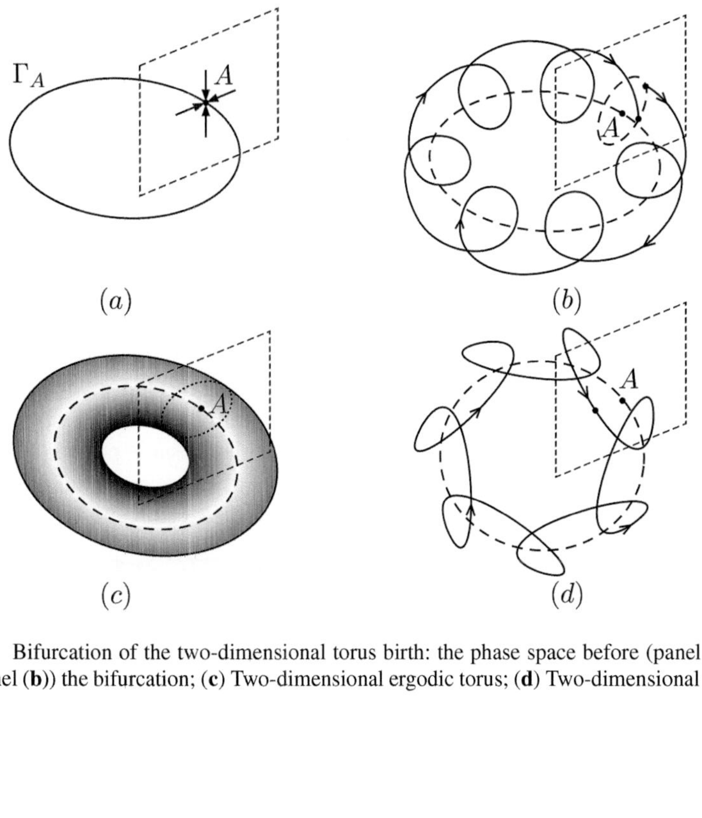

Fig. 1.22 Bifurcation of the two-dimensional torus birth: the phase space before (panel (**a**)) and after (panel (**b**)) the bifurcation; (**c**) Two-dimensional ergodic torus; (**d**) Two-dimensional resonant torus

Fig. 1.23 Symmetry-breaking bifurcation: the phase space before (panel (**a**)) and after (panel (**b**)) bifurcation

an invariant closed curve appears (Fig. 1.22b). A Neimark-Sacker bifurcation is realized in the Poincaré section. In the full phase space a stable **two-dimensional torus** T is born from the stable limit cycle Γ_A at the bifurcation moment. After that the limit cycle becomes unstable. In a case of irrational values $\omega/2\pi$ a trajectory on the torus never closes. It traces a surface of an ergodic torus (Fig. 1.22c). In a case of rational values $\omega/2\pi$ a resonance phenomenon occurs and a torus is said to be resonant (Fig. 1.22d). The resonant torus intersects the Poincaré section in a finite number of points.

One can realize a hard scenario of the torus birth. It corresponds to a situation when an unstable two-dimensional torus collides with a stable limit cycle and then disappears. After that the limit cycle becomes unstable and the possibility to observe new stable oscillatory regimes depends on global properties of a DS.

Symmetry-Breaking Bifurcation

The symmetry-Breaking Bifurcation can be observed in DSs with symmetric properties or in systems consisting of two or more identical partial subsystems. There exists a symmetric stable limit cycle in the phase space before the bifurcation (Fig. 1.23a). The bifurcation corresponds to a real multiplier passing through the value $\mu_i = 1$. In contrast to the saddle-node bifurcation, the initially stable limit cycle does not disappear after the symmetry-breaking bifurcation, but becomes saddle. It gives rise to birth of two stable cycles with the same period (Fig. 1.23b). New limit cycles are mutually symmetric, but are not symmetric in themselves. Due to this fact the bifurcation got its name. The described bifurcational transition of limit cycles corresponds to transition from monostability to bistability and in this sense is similar to a pitchfork bifurcation of steady states.

1.6.3 Nonlocal Bifurcations

Nonlocal bifurcations involve manifolds of saddle limit sets in the phase space. Basic cases of such bifurcations are considered below.

Separatrix Loop Formation

The simplest realization of separatrix loop formation is illustrated in Fig. 1.24 on the example of a saddle point on the phase plane. Here the saddle equilibria A has the stable manifold Θ_S and the unstable one Θ_U (Fig. 1.24a). Growth of a parameter m results in approach of manifolds Θ_S and Θ_U. At the bifurcation moment at $m = m^*$ the manifolds touch each other and form a closed phase trajectory (Fig. 1.24b) called a **separatrix loop of a saddle point**. The separatrix loop is nonrobust in a case of a dissipative DS and disappears at $m \neq m^*$. The behaviour of a DS after passing through the bifurcation value $m = m^*$ depends on the eigenvalues of the saddle point A: $\lambda_{1,2}(m^*)$. Then either situation can be observed:

- $\lambda_1(m^*)+\lambda_2(m^*) < 0$: The unstable manifold Θ_U is twisted inside. Then a stable limit cycle is born (Fig. 1.24c).
- $\lambda_1(m^*)+\lambda_2(m^*) > 0$: The unstable manifold Θ_U is twisted, but becomes outside of the stable manifold Θ_S. Then an unstable limit cycle is born (Fig. 1.24d).

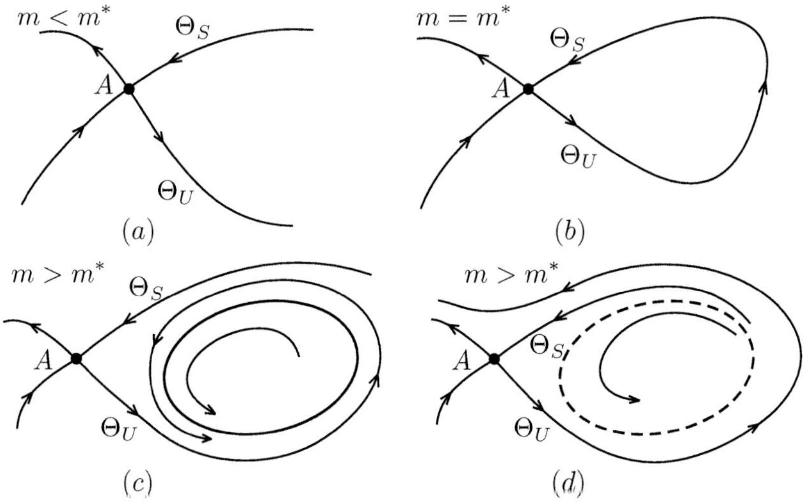

Fig. 1.24 Separatrix loop formation: the phase space before the bifurcation (panel (**a**)) and at the bifurcation moment (panel (**b**)); the phase space after the bifurcation for $\lambda_1(m^*) + \lambda_2(m^*) < 0$ (panel (**c**)) and $\lambda_1(m^*) + \lambda_2(m^*) > 0$ (panel (**d**)). The solid closed curve in panel (**c**) is a stable limit cycle, the dashed closed curve in panel (**d**) is an unstable limit cycle

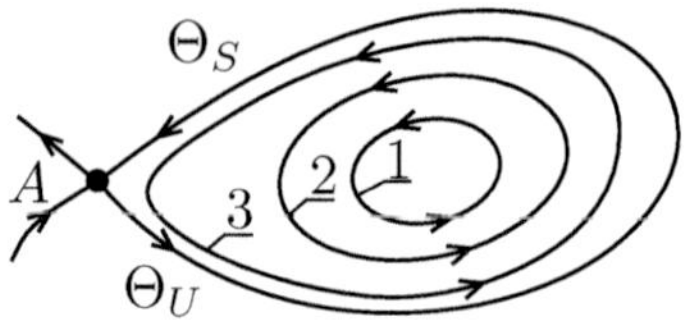

Fig. 1.25 Expansion of a stable limit cycle caused by parameter growth ($m_1 < m_2 < m_3$)

In the three-dimensional phase space separatrix loop formation can be realized in a more complicated way. Particularly, one must consider the sum $\lambda_1(m^*)+\lambda_2(m^*)+\lambda_3(m^*)$ in a case of a point of equilibrium 'saddle-node'. Similarly, in a case of an equilibrium point 'saddle-focus' characterized by three eigenvalues $\lambda_{1,2,3}$, such that λ_1 is real and $\lambda_{2,3}$ are complex conjugate quantities, the summary quantity dictating the bifurcation character becomes $\lambda_1(m^*) + \text{Re}(\lambda_2(m^*)) + \text{Re}(\lambda_3(m^*))$. Depending on the sign of the sum either birth of a limit cycle or more complex transitions are possible. So, bifurcation transition resulting in formation of a separatrix loop of a saddle-focus equilibrium can induce the chaotic dynamics (see Shilnikov's theorem).

Collision of a Separatrix Loop and a Stable Limit Cycle

This bifurcation is illustrated in Fig. 1.25. Increasing a system parameter m gives rise to expansion of a stable limit cycle. The limit cycle becomes closer and closer to a saddle equilibrium point and to an unstable separatrix contour. As a result, the period of self-oscillations increases because of slow motions in a neighbourhood of the saddle. At the bifurcation moment the period becomes infinite, the limit cycle collides with the separatrix loop and disappears. During this process the multipliers of the limit cycle tend to zero (if the limit cycle was initially stable). The dynamics after the bifurcation depends on the global properties of a DS.

Saddle-Node Bifurcation on a Separatrix Loop

This bifurcation consists in merging two equilibria: a saddle and a stable node (points A_S and A_N in Fig. 1.26a) into one nonrobust 'saddle-node' type equilibrium point (point A_{SN} in Fig. 1.26b). A separatrix loop existing before the bifurcation is formed by an unstable manifold of the saddle A_S closing at the stable node A_N. After the bifurcation the saddle-node disappears as well as the separatrix loop, and a stable limit cycle arises on the separatrix loop place (Fig. 1.26c).

Inverse varying a parameter (from Fig. 1.26c to a) gives rise to arising a saddle-node equilibrium on a stable limit cycle. Then one observes division of the saddle-node into a saddle and a stable node and the disappearance of a stable limit cycle. At the bifurcation moment the period of self-oscillations becomes infinite, the multipliers are zeros.

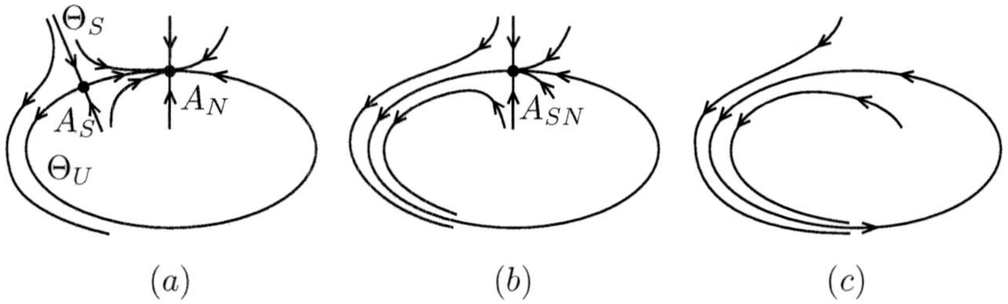

Fig. 1.26 Saddle-node bifurcation on a separatrix closed loop: before the bifurcation (panel (**a**)), at the bifurcation moment (panel (**b**)) and after the bifurcation (panel (**c**)). The closed curve in panel (**c**) is a stable limit cycle

References

1. S. Strogatz, *Nonlinear Dynamics and Chaos*, 2nd edn. (Westview Press, Boulder, 2014)
2. V. Anishchenko, V. Astakhov, A. Neiman, T. Vadivasova, L. Schimansky-Geier, *Nonlinear Dynamics of Chaotic and Stochastic Systems* (Springer, Berlin, 2007)
3. V. Anishchenko, T. Vadivasova, G. Strelkova, *Deterministic Nonlinear Systems* (Springer, Berlin, 2014)
4. G. Layek, *An Introduction to Dynamical Systems and Chaos* (Springer, Berlin, 2015)
5. E. Izhikevich, *Dynamical Systems in Neuroscience* (The MIT Press, Cambridge, 2007)
6. L. Arnold, *Random Dynamical Systems* (Springer, Berlin, 2003)

Chapter 2
Electronic Modelling: Classical Approach

Methods of electronic model synthesis used before the appearance of analog modelling are generally considered in the second chapter. All the described approaches and circuit solutions are described in more details on concrete examples in the fifth chapter. For the convenience of readers this chapter begins from basic formulations and definitions. The description is mostly focused on terms, definitions and laws used in the further chapters. Books [1–3] are recommended, if additional study is necessary.

2.1 Basic Terms and Principles

2.1.1 Current, Voltage, Power

An electric current is a directed flow of electrically charged particles. The motion is caused by an applied electric field or by an imbalance between positive and negative charges. Thus, the existence of the electric current requires two satisfied conditions:

- The presence of free charge carriers;
- The presence of an electric field inducing and supporting the electric current.

The particles transmit a positive or negative charge. The simplest example of a negatively charged particle is an electron. The positive charge can be transmitted by positive ions or holes. Usually two quantities are used (and measured in physical experiments) for the description of oscillatory processes in electronic circuits: a current (in most cases is marked by the symbol I) and a voltage (in most cases is marked by the symbol V).

The quantity **electric current** (or **current**) is the rate of flow of electric charge past a point or region (see, for instance, the section S in Fig. 2.1). In the simplest

© The Author(s), under exclusive license to Springer Nature Switzerland AG 2024

V. Semenov, *Electronic Modelling of Deterministic and Stochastic Oscillators*, Springer Series in Synergetics, https://doi.org/10.1007/978-3-031-81553-9_2

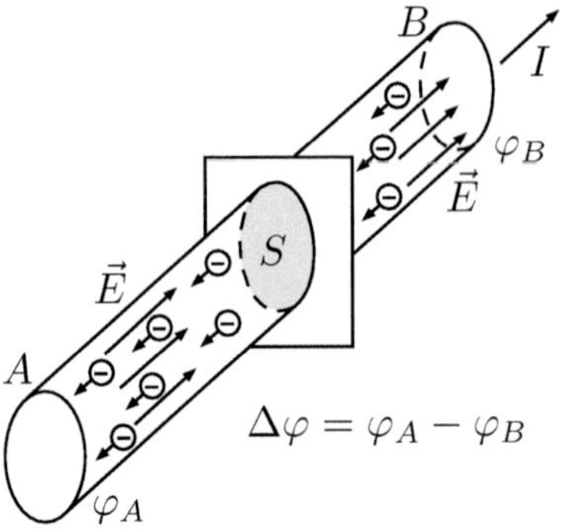

Fig. 2.1 Electric field lines and moved negative ions in a conductor with the potential difference $\Delta\varphi$ between poles A and B

case (steady flow of charge across a section of a conductor) the current is calculated as $I = Q/t$, where Q is the electric charge transferred through the surface over a time t. More generally, the electric current can be represented as the time derivative of the charge Q: $I = dQ/dt$. The SI unit of the electric current is the ampere marked by the symbol A. The current 1 ampere ($I = 1$ [A]) corresponds to the electric charge passing across a surface at the rate of one coulomb per second.

The quantity **voltage** is defined as the work of the electric field per unit of charge, which is necessary for motion of a test charge between two points. Generally, the motion is induced by contributions of two works: the work of an electric field caused by a potential difference between two points or poles (the potential difference $\Delta\varphi = \varphi_A - \varphi_B$ in Fig. 2.1) and the work of external forces. The second work is characterized by the quantity ε_{ext} called an **electromotive force**. In summary, a voltage value becomes $V = \varphi_A - \varphi_B + \varepsilon_{\text{ext}}$. If the electromotive force is absent, then the voltage fully coincides with the potential difference $V = \varphi_A - \varphi_B$. The SI unit of the voltage is the volt marked by the symbol V. The voltage one volt ($V = 1$ [V]) corresponds to one joule of work spent for moving the positive charge 1 coulomb from a more negative point (lower potential) to a more positive point (higher potential).[1]

It is important to note the following difference: currents flow through elements, and voltages are applied across elements. Then the voltage is measured between two points of a circuit, but the current is measured in one point of a device or a connection. In addition, the formulation 'the voltage at a point in a circuit' is frequently used. This phrase implies the voltage between that point and the 'ground', a common reference point in a circuit considered to have zero potential. A 'signal ground', a 'chassis ground' and an 'earth ground' are distinguished. The last one marked by the symbol in Fig. 2.2a is mentioned in the further circuits in pages of this book. Using the 'earth ground', one considers literally the potential of the ground (i.e., earth or soil) as a reference.

[1] There also exist other definitions of the unit volt associated with electric power, or resistance, or a force of charged-body interaction.

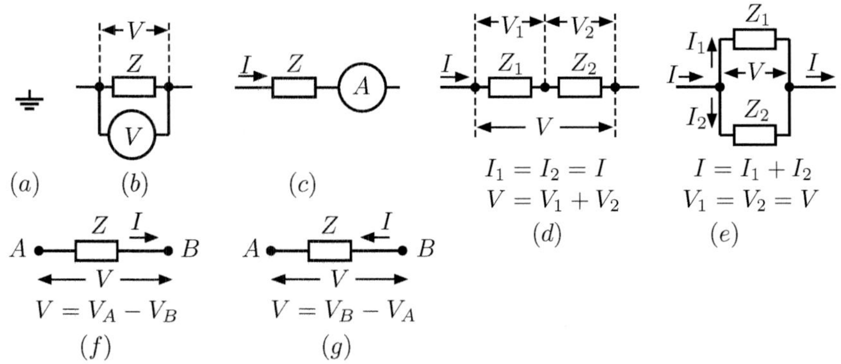

Fig. 2.2 Circuit depiction of a 'ground' point (panel (**a**)); Circuit connection of a voltmeter (panel (**b**)) and an ammeter (panel (**c**)); Two circuit elements coupled in series (panel (**d**)) and in parallel (panel (**e**)) and related formulas for currents and voltages. Panels (**f**) and (**g**) illustrate the sign convention for voltage polarity and current direction in two-terminal elements

Some of tools used for measure of the voltage are a voltmeter and an oscilloscope connected between two points of the circuit (Fig. 2.2b). In the simplest case the electric current is measured using a device called an ammeter. To measure the current you have to cut a circuit and to install the ammeter there (Fig. 2.2c).

In a case of a direct current circuit (DC circuit) currents and voltages are non-variable. Then a voltmeter and an ammeter switched to the DC-regime show an instantaneous value of currents and voltages being constant. In a case of an alternating current circuit (AC circuit) the voltages and currents are time-varying quantities and change according to the sinusoidal law: $I(t) = I_{\text{peak}} \sin(\omega t)$ and $V(t) = V_{\text{peak}} \sin(\omega t)$. In most cases the voltmeter and the ammeter switched to the AC-regime indicate the RMS-value.[2] The relationship between the peak-value and the RMS-average for sinusoidal voltages and currents is: $V_{\text{RMS}} = V_{\text{peak}}/\sqrt{2}$ and $I_{\text{RMS}} = I_{\text{peak}}/\sqrt{2}$.

There are two general rules concerning voltages and currents:

- The sum of input currents equals the sum of output currents at any point of a circuit;
- Elements connected in parallel have the same voltage across them.

It leads from the first rule that the same current flows through elements coupled in series (Fig. 2.2d). In such a case the summary 'voltage drop' equals to the sum of particular 'voltage drops' across each element. The first rule also means that the

[2] The root mean square value of a sinusoidally-varying quantity is the square root of the mean value of the quantity squared values taken over an oscillation period T. Mathematically, it is written as being: $x_{\text{RMS}} = \sqrt{\frac{1}{T}\int\limits_0^T x^2 dt}$.

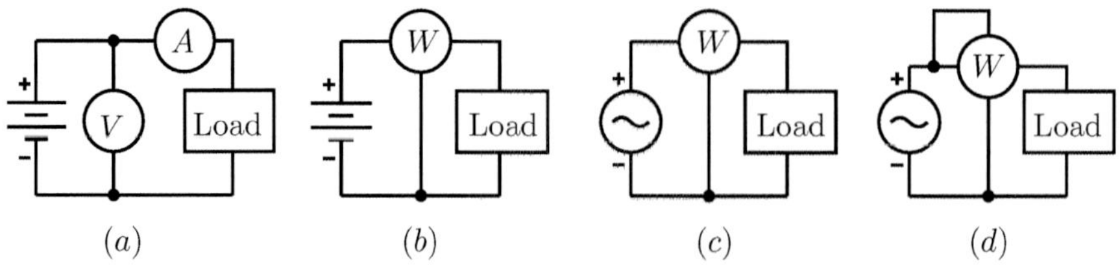

Fig. 2.3 Measurement of electric power from a DC-power source using a voltmeter and an ammeter (panel (**a**)) and a single wattmeter (panel (**b**)); Measurement of electric power consumed from an AC-power source; (**d**) Four-terminal depiction of a wattmeter

current flowing through a circuit made up of elements coupled in parallel equals to the sum of currents passing through the elements (Fig. 2.2e).

Two-terminal elements (resistors, inductors, etc.) are described by functional relationships between currents and voltages (Ohm's law, Faraday's law, etc.). To correctly use the relationships and to derive equations describing electronic circuits one needs to choose current directions and to determine the signs of the corresponding potential differences across two-terminal elements. Voltage polarity and current direction have to be consistent with each other. There exists the universal sign convention which allows to specify the sign relationships and to use them for the equation derivations (compare panels (f) and (g) in Fig. 2.2). This approach is used in the further derivations of oscillatory circuit equations.

An important quantity characterizing electrical systems and different circuit elements is **electric power**. The SI unit of the electric power is watt market by the symbol W. In a case of DC circuits the electric power consumed by a circuit element is calculated as $P = VI$. For example, the current passing through an electric device consuming the power $P = 45$ [W] at the applied voltage $V = 15$ [V] becomes $I = P/V = 3$ [A]. In a case of AC circuits one distinguishes the instantaneous power given by the formula $P_{\text{ins}}(t) = V(t)I(t)$ and the average power $P_{\text{avg}} = V_{\text{RMS}} I_{\text{RMS}} \cos(\phi)$, where ϕ is the phase difference between the current and the voltage.

In the simplest case the electric power consumed by a load from a DC-power source (for example, it can be a battery, a laboratory DC-power source, etc.) can be measured using a voltmeter and an ammeter switched to the DC-regime (Fig. 2.3a). Then one calculates $P = VI$ assumed to be constant. The second way is to use a wattmeter switched to the DC-regime (Fig. 2.3b).

If a load consumes electric power from an AC-power source (it can be a generator, an electrical outlet, a transistor-based amplifier, etc.), one can similarly use the voltmeter and the ammeter switched to the AC-regime, but only if the phase difference between the current and the voltage is known. Then the average power P_{avg} can be calculated. If the phase difference is unknown, then it is necessary to use the wattmeter switched to the AC-regime (Fig. 2.3c). In electronic circuits the wattmeter is schematically depicted using three terminals (Fig. 2.3a–c) or four terminals (Fig. 2.3d).

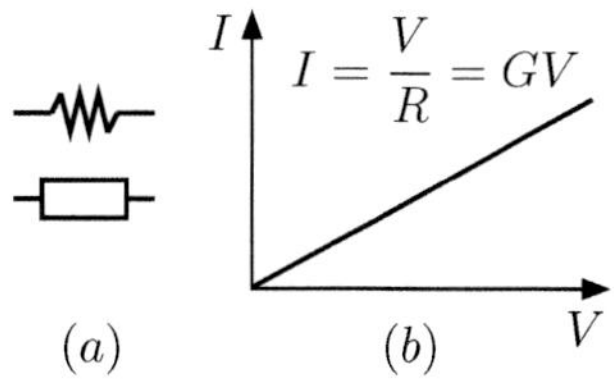

Fig. 2.4 Depiction of a resistor (panel (**a**)) and its current-voltage characteristic (panel (**b**))

2.1.2 Resistor and Ohm's Law

It has been empirically highlighted that the current through metallic conductors is proportional to the voltage across them. This relationship was first established by German scientist Georg Simon Ohm and written as the formula $I = V/R$ called **Ohm's law**. The quantity R is called a **resistance**. The larger is the resistance, the smaller is the current. The resistance is measured in ohms $[\Omega]$. The conductor resistance 1 ohm ($R = 1\ [\Omega]$) corresponds to the applied voltage of one volt producing the current of one ampere through the conductor. Depending on a concrete situation, the resistance of circuit elements is neglected (for instance, the resistance of wire conductors used in circuits is usually assumed to be zero) or taken into consideration.

A two-terminal circuit element being a conductor with constant resistance is a **resistor**. The resistor is made out of some conducting materials, with a wire or contacts at each terminal. Depictions of the resistor are shown in Fig. 2.4a. **The current-voltage characteristic**[3] of the resistor represents a linear dependence with a coefficient $k = 1/R$ (Fig. 2.4b).

Ohm's law can be rewritten in the form $I = GV$. The quantity G is called a **conductance**. The larger is the conductance, the larger is the current. The resistor conductance is the reciprocal of the resistance. The electrical conductance SI unit is siemens $[S] = [\Omega^{-1}]$. The electric current through a device with a conductance of one siemens will increase by one ampere for every increase of one volt of the voltage across the device.

An ideal resistor strictly follows Ohm's law for any voltage and current values. Its resistance does not depend on external conditions (for instance, on a temperature). The ideal resistor has no reactive properties (an ideal resistor operating inside AC-circuits has zero phase difference between the current and the voltage).

Since the same current I flows through two resistors connected in series (Fig. 2.5a), one can write $V = V_1 + V_2 = IR_1 + IR_2$. Then the summary resistance $R = V/I$ becomes $R = R_1 + R_2$. The summary conductance value $G = I/V$ can be derived using the equalities $G = 1/R$, $G_1 = 1/R_1$, and $G_2 = 1/R_2$. Then one obtains $1/G = 1/G_1 + 1/G_2$ or $G = \frac{G_1 G_2}{G_1 + G_2}$.

[3] A current-voltage characteristic (I-V characteristic) is a curve illustrating the dependence of the current through a circuit element on the voltage across it. The dependence in Fig. 2.4b is obtained by application direct voltages (output voltages of DC-power sources) across the element.

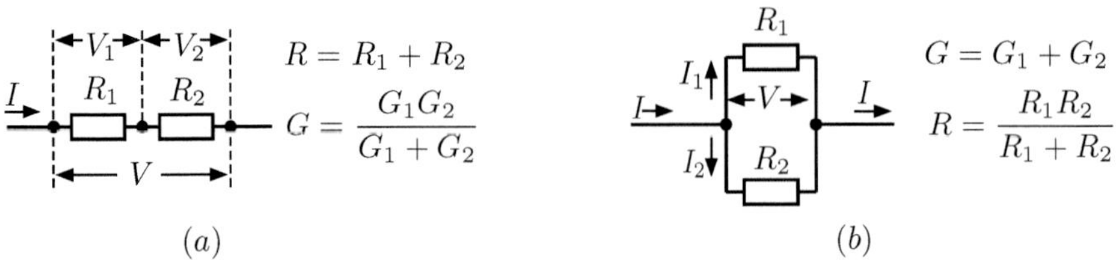

Fig. 2.5 Two resistors connected in series (panel (**a**)) and in parallel (panel (**b**)) and corresponding relationships for resistances and conductances

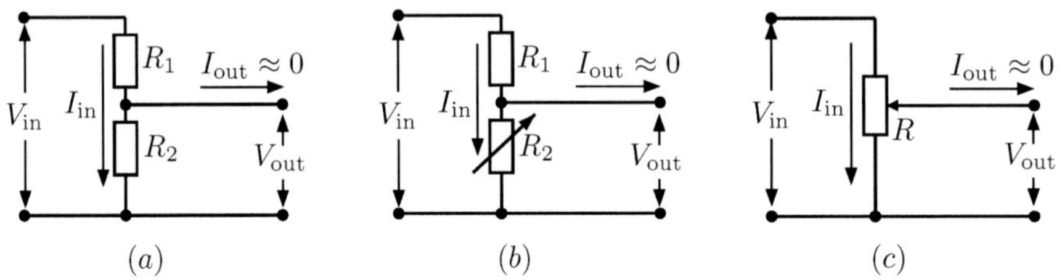

Fig. 2.6 Voltage divider with two fixed resistors (panel (**a**)), with an adjustable resistor R_2 (panel (**b**)) and a single potentiometer (panel (**c**))

Similarly, the voltage drops across two resistors connected in parallel are the same. Then the following equality is right: $I = I_1 + I_2 = VG_1 + VG_2$. The summary conductance $G = I/V$ becomes $G = G_1 + G_2$. Then the corresponding summary resistance $R = 1/G$ can be derived, $R = \frac{R_1 R_2}{R_1 + R_2}$.

2.1.3 *Voltage Divider*

To put it briefly, a voltage divider is a circuit that produces an output voltage being a fraction of the input one. The simplest voltage divider is shown in Fig. 2.6a. It consists of two resistors R_1 and R_2. It is assumed that there is no load on the output (output current I_{out} is neglected). In such a case the same current $I_{\text{in}} = V_{\text{in}}/(R_1 + R_2)$ flows through the resistors $R_{1,2}$. Then the output voltage across the resistor R_2 becomes $V_{\text{out}} = IR_2 = \frac{R_2}{R_1+R_2} V_{\text{in}}$. It is important to note that $V_{\text{out}} \le V_{\text{in}}$. For this reason the circuit is called a divider.

The voltage divider is a simple but frequently used circuit. It will be included into the circuits considered in this book. If either resistor is adjustable (for example, R_2 in Fig. 2.6b), then one can control and vary the output voltage. Sometimes it is much easier to use a single variable resistor, or potentiometer, instead of two resistors (Fig. 2.6c).

2.1.4 Inductor and Faraday's Law

An inductor, also called a coil, is a passive two-terminal element which typically consists of an insulated wire wound into a coil. Typical inductor depictions are shown in Fig. 2.7a. The simplest case will be considered when the inductor has no core. When an electric current flows through an inductor, it induces a magnetic field around it. The generated magnetic flux linkage Φ is proportional to the current: $\Phi = LI$. The quantity L is called an **inductance**. The inductor operates according to Faraday's law of induction: any flux change (or any current change) induces an electromotive force in the inductor being proportional to the rate of flux change: $V_L = -\frac{d\Phi}{dt}$. The negative sign in the equation means that the induced electromotive force strives to reduce the current change that created it. If the current increases, the voltage and tends to reduce the current. If the current is decreasing, the voltage tends to maintain the current. Then physical meaning of inductance becomes visible: L is the ratio between the induced voltage and the rate of change of the current: $V = L\frac{dI}{dt}$. The SI unit of the inductance in the is henry [H]. One henry is the amount of the inductance which generates a voltage of one volt when the current is changing at a rate of one ampere per second.

An ideal inductor strictly follows the Faraday's law for any voltage and current values. Its inductance does not depend on external conditions (for instance, on a temperature) and remains to be fixed. The ideal inductor has no resistive properties (the inductor resistance R_L equals to zero).

Since the same current I flows through two inductors connected in series (Fig. 2.7b), one can write $V = V_1 + V_2 = L_1\frac{dI}{dt} + L_2\frac{dI}{dt} = (L_1 + L_2)\frac{dI}{dt}$. Then the summary inductance becomes $L = L_1 + L_2$.

Similarly, the voltage drops across two inductors connected in parallel are the same (see Fig. 2.7c). Then the following equality is right: $\frac{dI}{dt} = \frac{dI_1}{dt} + \frac{dI_2}{dt} = \frac{1}{L_1}V + \frac{1}{L_2}V = \left(\frac{1}{L_1} + \frac{1}{L_2}\right)V = \frac{1}{L}V$, where L is the summary inductance calculated as $\frac{1}{L} = \frac{1}{L_1} + \frac{1}{L_2}$ or $L = \frac{L_1 L_2}{L_1 + L_2}$.

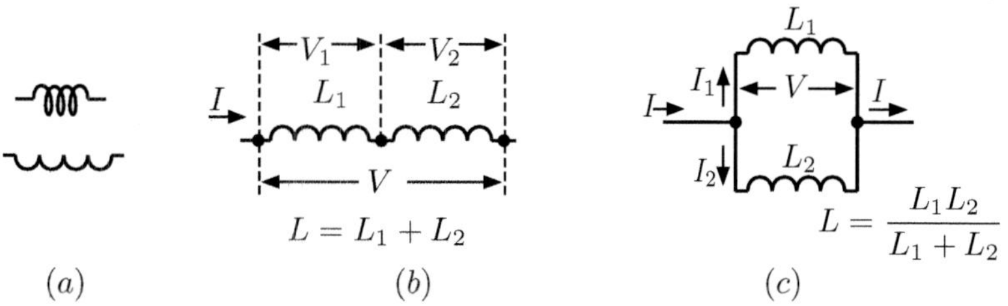

Fig. 2.7 (**a**) Two options for depiction of an inductor; Two inductors connected in series (panel (**b**)) and in parallel (panel (**c**)) and corresponding relationships for inductances

2.1.5 Capacitor

The capacitor depiction is shown in Fig. 2.8a. Basically, a capacitor can be imagined as two conducting plates separated by a non-conducting layer (for instance, vacuum, air, any dielectric material layer, etc.). Both plates have electrodes being two terminals of the capacitor. According to Coulomb's law, a charge on one conductor exerts a force on the charge carriers within the other conductor. The force attracts opposite-polarity charge and repels like-polarity charges. An electrical charge on either plate of the capacitor induces the opposite charge on the surface of the other plate. It gives rise to an electric field arising in the dielectric layer. If the separation between the plates is assumed to be uniform over the plate area, the distance between the plates is small compared to the plate dimensions, and boundary effects are neglected, then the electric field anywhere between the plates $\vec{E}$ is constant and directed perpendicularly to the plate surface. Thus, if a positive charge $+Q$ is placed on either plate and a negative charge $-Q$ on the other one, then the charge on each plate is uniformly spread on the inside surface of each plate with the constant charge density $\sigma = \pm Q/S$ (where S is the square of each plate) coulombs per square meter. The corresponding electric field strength becomes $E = \sigma/\varepsilon$, where ε is the permittivity of the middle dielectric layer. Finally, the potential difference V between the plates defined as the line integral of the electric field over a line from one plate to another equals to $V = \frac{Qd}{\varepsilon S}$ or $V = Q/C$. The quantity C called a **capacitance** relates the voltage V and the charge Q. Differentiating the linear relationship $V = Q/C$ over time, one obtains: $\frac{dV}{dt} = \frac{1}{C}\frac{dQ}{dt}$ or $I = C\frac{dV}{dt}$, where I is a current through the capacitor.

The SI capacitance unit is the farad [F]. The capacitance of one farad means that one coulomb of charge on each plate surface causes the voltage of one volt across the device. An ideal capacitor is characterized by a constant capacitance C and has no insulation current.

Since the same current I flows through two capacitors connected in series (Fig. 2.8b), one can write $V = V_1 + V_2 = \frac{1}{C_1}\int_0^t I dt + \frac{1}{C_2}\int_0^t I dt$. After differentiation

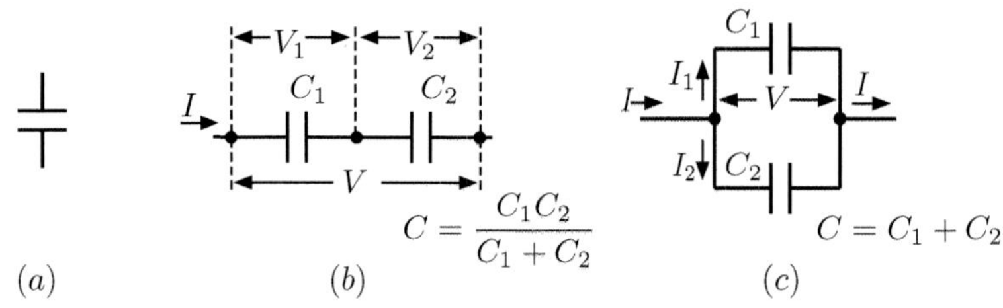

Fig. 2.8 (**a**) Depiction of a capacitor; Two capacitors connected in series (panel (**b**)) and in parallel (panel (**c**)) and corresponding relationships for capacitances

one obtains: $\frac{dV}{dt} = \left(\frac{1}{C_1} + \frac{1}{C_2}\right) I$. Then the summary capacitance C obeys the equality $\frac{1}{C} = \frac{1}{C_1} + \frac{1}{C_2}$.

Similarly, the voltage drops across two capacitors connected in parallel are the same (Fig. 2.8c). Then the following equality is right: $I = I_1 + I_2 = C_1\frac{dV}{dt} + C_2\frac{dV}{dt} = (C_1+C_2)\frac{dV}{dt}$. That means the summary capacitance becomes $C = C_1+C_2$.

2.1.6 Complex Current, Voltage and Electrical Impedance

The ideal resistor model implies that the element connected into a DC-circuit operates in exactly the same way as in an AC-circuit at any frequency: the ideal resistor strictly follows Ohm's law. If the applied voltage is sinusoidal as in Fig. 2.9a, $V_R = V_{\max}\cos(\omega t)$, then the flowing current becomes $I_R = \frac{V_{\max}}{R}\cos(\omega t)$ and the phase shift between the current and the voltage is zero.

Current-voltage relationships of a capacitor and an inductor (known collectively as reactive components) connected in AC-circuits are more complicated and contain a dependence on a frequency. Indeed, if the ideal capacitor is connected to an AC-voltage source (Fig. 2.9b) and the applied voltage is sinusoidal, $V_C = V_{\max}\cos(\omega t)$, then the flowing current is $I_C = C\frac{dV}{dt} = \omega C V_{\max}(-\sin(\omega t)) = \omega C V_{\max}\cos(\omega t + \frac{\pi}{2})$. That means the current leads the voltage with the phase shift $\pi/2$.

Similarly, if the ideal inductor is connected to an AC-voltage source (Fig. 2.9c) and the applied voltage is sinusoidal, $V_L = V_{\max}\cos(\omega t)$, then the flowing current

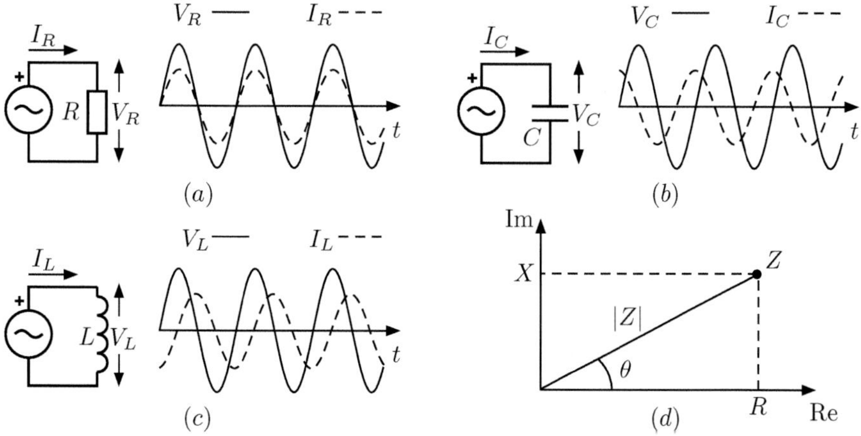

Fig. 2.9 Ideal resistor (panel (**a**)), capacitor (panel (**b**)) and inductor (panel (**c**)) connected to an AC-voltage source and schematically depicted time realizations of the applied voltages and the flowing currents; (**d**) Graphical impedance representation on the complex plane

is $I_L = \frac{1}{L}\int_0^t V_L dt = \frac{V_{\max}}{\omega L}\cos(\omega t - \frac{\pi}{2})$. That means the current lags the voltage with the phase shift $\pi/2$.

To simplify calculations, sinusoidal voltage and current signals are commonly represented as complex-value functions of time: $V = |V|\exp(j(\omega t + \phi_V))$, $I = |I|\exp(j(\omega t + \phi_I))$, where j is the imaginary unit. Characterizing two-terminal elements and circuits in the framework of the complex representation, one introduces a quantity Z called an **impedance** defined as the ratio of the complex voltage and the complex current, $Z = \frac{V}{I} = \frac{|V|}{|I|}\exp(\phi_V - \phi_I)$. As you can see, the impedance is an extension of a resistance to AC-circuits. In contrast to the resistance, the impedance simultaneously relates both magnitudes and phases of currents and voltages. In a case of DC-circuits there is no difference between the impedance and the resistance since phases are out of consideration and the resistance can be considered as impedance with zero phase angle. The SI unit for resistance and impedance is the same, the ohm [Ω]. The reciprocal of the impedance is a complex admittance, whose SI unit is the siemens. In the context of multiple port networks the impedance quantity concept is extended to the impedance matrix which relates input and output voltages and currents.

Impedance is a complex number and can be presented in the polar form, $Z = |Z|\exp(j\theta)$ and in Cartesian form, $Z = R + jX$, where the real part R is the resistance and the imaginary part X is the reactance. Both representations are depicted in Fig. 2.9d. The polar form of the complex impedance has the following physical sense:

- The magnitude of the complex impedance, $|Z| = |V|/|I|$, is the ratio of the voltage amplitude to the current amplitude (here the magnitude equation $|V| = |Z||I|$ is the familiar Ohm's law);
- The phase of the complex impedance, $\theta = \phi_V - \phi_I$, is the phase shift by which the current lags the voltage.

The total impedance of two-terminal elements with the particular impedances $Z_1, Z_2, \ldots, Z_n$ connected in series becomes: $Z = Z_1 + Z_2 + \ldots + Z_n$. The total impedance of the same elements connected in parallel is calculated using the formula: $\frac{1}{Z} = \frac{1}{Z_1} + \frac{1}{Z_2} + \ldots + \frac{1}{Z_n}$. The total impedance formulas are derived using the same principles as for the total resistance derivations in Sect. 2.1.2.

Using the impedance, one can analytically derive currents and voltages in the complex form. Real currents $I = I_{\max}\cos(\omega t + \phi_I)$ and voltages $V = V_{\max}\cos(\omega t + \phi_V)$ measured in physical experiments are real parts of the corresponding complex quantities (consequence of Euler's formula).

Analyzing the discussed above current-voltage dependences for the resistor, capacitor and inductor connected to AC-voltage sources, one can derive the impedances of the elements: $Z_R = R$, $Z_C = \frac{1}{j\omega C}$, $Z_L = j\omega L$.

2.1.7 Nonlinear Two-Terminal Elements

Discussed above models of a resistor, a capacitor and an inductor imply validity of the linear relationships $I = V/R$, $V = Q/C$, $\Phi = LI$. The parameters R, C and L are assumed to be fixed. However, real two-terminal elements can have variable or state-dependent properties and the corresponding relationships become nonlinear. For example, increase of a current through a resistor can lead to grow of the resistor temperature. Then the resistance increases. In such a case one can describe the resistor properties by a resistance functional dependence on a temperature $R = R(T)$ or a flowing current $R = R(I)$. Similarly, the inductance and the capacitance can be current- and voltage-dependent, especially in a case of high voltages and currents. Moreover, properties of capacitors and inductors connected into AC-circuits can be frequency-dependent.

There exist a broad spectrum of resistive two-terminal elements, which are principally nonlinear. Their properties are reflected in nonlinear current-voltage characteristics obtained by changing a current or a voltage in a DC-regime. For example, one of such elements is a varistor whose resistance depends on an applied voltage (Fig. 2.10a). In addition, properties of elements based on p-n junction are strongly nonlinear and their current-voltage curves can be monotonic (as the rectifier diode characteristics in Fig. 2.10b) or non-monotonic (as the tunnel diode characteristics in Fig. 2.10c).

Nonlinearity can take more complicated forms, such as in memristive elements [4, 5]. A class of the memristive systems is identified by a continuous functional dependence of characteristics at any time on previous states. Particularly, there are many elements, which exhibit the memristor properties. The memristor resistive properties depend on the entire past history of a voltage or a current. In the simplest case, the memristor current-voltage dependence is described by Ohm's law $V = R(z)I$, where z is a state variable which varies according to the state equation $\frac{dz}{dt} = f\left(\int\limits_{-\infty}^{t} V(t)dt\right)$ or $\frac{dz}{dt} = f\left(\int\limits_{-\infty}^{t} I(t)dt\right)$. The first configuration corresponds to a flux-controlled memristor, while the second one describes a charge-controlled

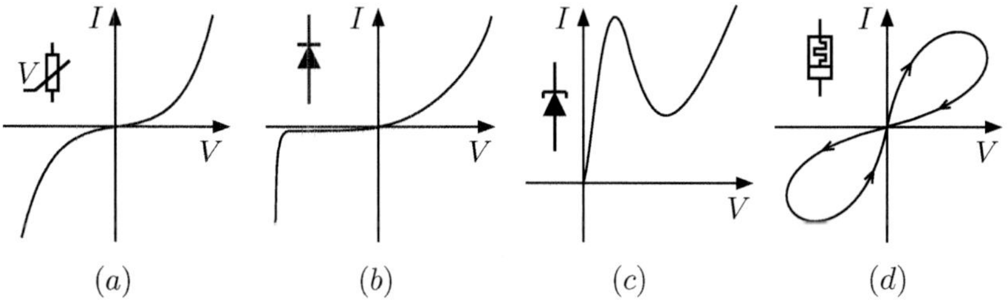

Fig. 2.10 Typical current-voltage characteristics of a varistor (panel (**a**)), a rectifier diode (panel (**b**)), a tunnel diode (panel (**c**)) connected with a DC-voltage source and a memristor driven by an AC-voltage source (panel (**d**))

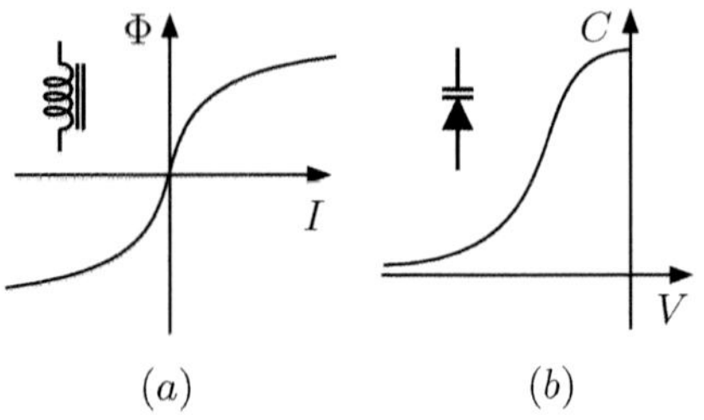

Fig. 2.11 (**a**) Flux-current dependence of an iron core inductor; (**b**) Volt-farad characteristics of a varicap diode

memristor. A memristor forced by the periodic voltage signal $V_{ext} = A_{ext}\sin(\omega t)$ (AC-regime) exhibits the current-voltage characteristic, which depicts a hysteresis loop being pinched with growth of the frequency ω (Fig. 2.10d).

The inductor nonlinear properties become more visible in a case of large values of an applied voltage or a flowing current as well as in the presence of a core. The range of air core inductance values is limited. In order to increase the inductance value, an iron core is placed inside an inductor. For the iron core inductor, the relationship between the magnetic flux $\Phi(t)$ and the current $I(t)$ passing through the inductor can be described by the cubic function: $I(t) = \frac{N\Phi(t)}{L_0} + A\Phi^3(t)$, where N is the number of turns of the coil, L_0 is the self-inductance of the coil, a parameter A characterizes the nonlinearity strength. The flux-current curve of such elements are made up of linear part, and two branches corresponding to saturation (Fig. 2.11a). Consequently, the properties of circuits including the nonlinear inductor can be significantly different from ones based on the linear inductor.

Similarly to real inductors, real capacitors demonstrate nonlinear features in a case of large applied voltage (of course, if they are able to operate in these regimes) or their dielectric isolator can exhibit specific intrinsic peculiarities. In such cases one deal with nonlinear dependence $Q(V)$ and the corresponding current-voltage relationship becomes: $I = C(V)\frac{dV}{dt}$. Furthermore, there exist semiconductor devices with nonlinear capacitance. One of them is a varicap diode whose capacitance is voltage-dependent and can be described by the function $C(V) = C_0\sqrt{\frac{V_0}{V_0 - V}}$ with condition $V \leq V_0$ (Fig. 2.11b).

2.1.8 *Two-Port Networks*

A two-port network is an electrical circuit or device with two pairs of terminals connected to external circuits. Two terminals constitute a port if the electric current entering one terminal equals to the current emerging from the other terminal of the same port. The schematically depicted in Fig. 2.12a two-port network has the input (port 1) and the output (port 2) ports. A two-port network is regarded as a 'black box' with its properties described by a matrix. This approach allows to easily

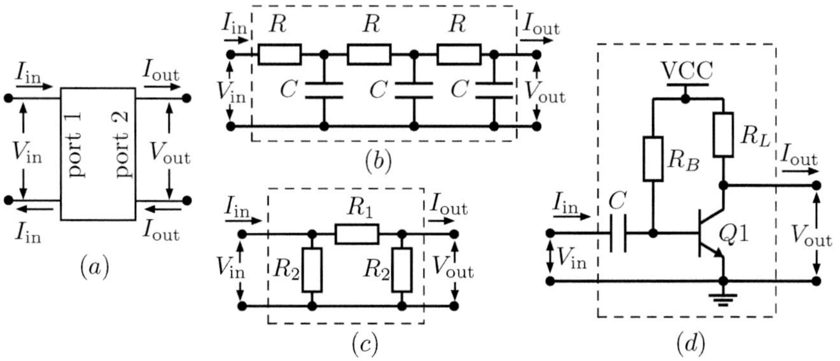

Fig. 2.12 Schematic depiction (panel (**a**)) and examples of a two-port network: a low frequency filter (panel (**b**)), an attenuator (panel (**c**)) and an amplifier based on a bipolar n-p-n transistor (panel (**d**))

calculate the network response to signals applied to the ports without additional calculations of internal voltages and currents in the network. Thus, one can describe by the same way a wide spectrum of devices such as filters (Fig. 2.12b), attenuators (Fig. 2.12c), amplifiers (Fig. 2.12d), signal converters, transmission lines, etc. The methodology implies the description of a two-port network by using a 2×2-square matrix of numbers (in the general case, the matrix consists of complex numbers). Depending on a concrete situation, matrices of z-parameters (impedance parameters), y-parameters (admittance parameters), h-parameters (hybrid parameters), g-parameters (inverse hybrid parameters) or ABCD-parameters (also known as chain, cascade, or transmission parameters) are used. The matrix elements establish relationships between the variables in Fig. 2.12a: V_{in} (the voltage across the port 1), I_{in} (the current into the port 1), V_{out} (the voltage across the port 2), I_{out} (the current into the port 2).

The most important two-port network parameters which are used in the further sections are the input resistance $R_{\text{in}} = V_{\text{in}}/I_{\text{in}}$, the output resistance $R_{\text{out}} = V_{\text{out}}/I_{\text{out}}$, the voltage amplification factor also called a gain $G = V_{\text{out}}/V_{\text{in}}$ and the transfer voltage ratio function $G = V_{\text{out}}/V_{\text{in}}$ in a case of signal transformers. All of them are real numbers. If a phase shift between currents and voltages in AC-circuits including two-port networks takes place, then one deals with the input impedance $Z_{\text{in}} = V_{\text{in}}/I_{\text{in}}$ and the output impedance $Z_{\text{out}} = V_{\text{out}}/I_{\text{out}}$ represented as complex quantities. The input resistance and impedance characterise the power consumption of a two-port network from an input signal source. The output resistance and impedance allow to determine the output current flowing through a load and the corresponding output power.

2.1.9 Kirchhoff's Laws

Kirchhoff's circuit laws (also called Kirchhoff's rules) represent two equalities for currents and potential differences (or voltages). **Kirchhoff's first law** states that the sum of currents flowing into any node in an electrical circuit is equal to the sum of currents flowing out of that note. That law can be equivalently reformulated: the algebraic sum of currents in a network of conductors meeting at a point is zero. Mathematically, it is written as $\sum_{i=1}^{n} I_i = 0$, where n is the total number of branches with currents flowing towards or away from the node. **Kirchhoff's second law** states that the sum of potential drops around a closed circuit loop equals to the sum of electromotive forces of sources included into the loop. Mathematically, it is written as being $\sum_{i=1}^{n} V_i = \sum_{j=1}^{m} \varepsilon_j$, where n is the total number of passive elements of the loop, m is the total number of included electromotive force sources. If the loop does not contain any sources of an electromotive force, then the sum of potential drops around the loop equals to zero.

Let us demonstrate the mentioned laws on the example of the circuit depicted in Fig. 2.13a. First of all, one must choose the directions of all current flows and traversals along each circuit loop assumed to be positive. You can choose any directions as you wish. If any real current in a circuit has the opposite direction, then after all calculations you just obtain a negative value of the corresponding current. After choosing the directions, Kirchhoff's current law is written for all nodes. A particular configuration of the chosen current directions and traversals along each loop is depicted in Fig. 2.13a. Then Kirchhoff's first law for the nodes B, D, G, H becomes: $I_1 = I_5 + I_2$, $I_2 = I_6 + I_3$, $I_3 = I_4 + I_1$, and $I_4 + I_5 + I_6 = 0$ correspondingly.

Next step is to write Kirchhoff's second law for voltage drops and electromotive forces. The voltages and the electromotive forces in the loop may be summed in

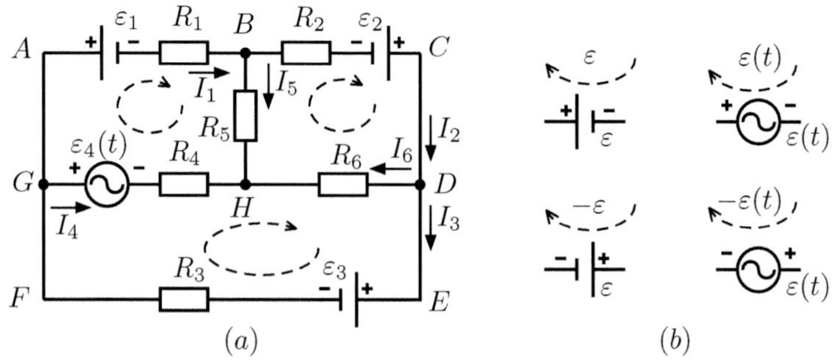

Fig. 2.13 (**a**) Example circuit for the explanation of Kirchhoff's laws; (**b**) Sign of an electromotive force produced by batteries (left panels) and functional generators (right panels) depending on the traversal direction

either direction. It makes no difference except of changing all the signs in the resulting equation. It does not matter where you start on the loop. If the direction of a current passing through a circuit branch coincides with the direction of traversal along the loop, then the corresponding voltage drop is assumed to be positive. Otherwise, the voltage drop is considered to be negative. Depending on polarity of an electromotive force source with respect to the direction of traversal along the loop (Fig. 2.13b), its electromotive force is taken into account with the sign '+' or '−'. Then Kirchhoff's voltage law for each contour in Fig. 2.13a takes the following form:

- Contour ABHGA: $I_1R_1 + I_5R_5 - I_4R_4 = \varepsilon_4(t) - \varepsilon_1$;
- Contour BCDHB: $I_2R_2 + I_6R_6 - I_5R_5 = \varepsilon_2$;
- Contour DEFGD: $I_4R_4 - I_6R_6 + I_3R_3 = -\varepsilon_3 - \varepsilon_4(t)$.

Simultaneous solution of equalities obtained from Kirchhoff's laws for currents and voltages allow to express and to find all the values of currents and voltage drops.

2.2 RLC-Circuit

A RLC-circuit is an electrical circuit made up of a resistor, an inductor, and a capacitor connected in series or in parallel. Such circuits have many applications. Furthermore, different modifications of RLC-circuits allow to realize a broad manifold of dynamical systems by means of electronic circuits. First of all, the RLC-circuits are considered below in the context of differential equations and mathematical models. A linear RLC-circuit has already discussed in Chap. 1 (Sect. 1.1). For this reason, the further consideration is switched to a nonlinear case.

2.2.1 *Nonlinear Series Oscillatory Circuit*

The simplest kind of a nonlinear series oscillatory circuit is depicted in Fig. 2.14a. It includes a nonlinear two-terminal element N_1. The voltage across the element N_1 is assumed to be nonlinearly dependent on the current passing through the element: $V_{N1} = f(I_{N1})$. The same current passes through the elements N_1,L and C. Kirchhoff's second law for the circuit in Fig. 2.14a is $V_{N1} + V_L + V_C = 0$, which can be rewritten for $V = V_C$ and I in the form $L\frac{dI}{dt} + V + f(I) = 0$. Taking into account the equality $I = C\frac{dV}{dt}$, one obtains the system of two first-order differential equations:

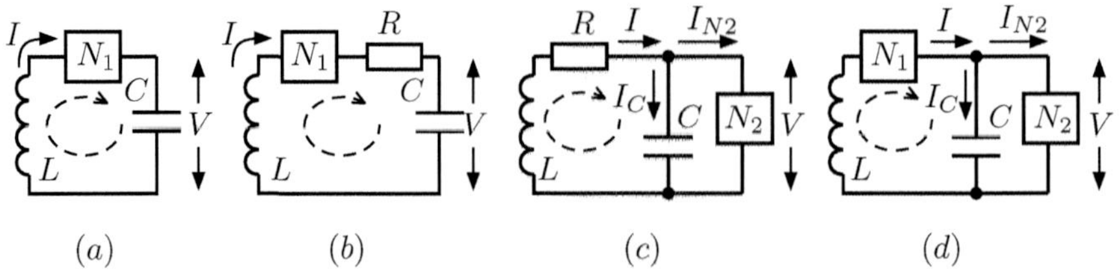

Fig. 2.14 Nonlinear series oscillatory circuit including (panel (**b**)) and not including (panel (**a**)) an additional resistor; (**c**) Linear RLC-circuit with a nonlinear element connected in parallel with a capacitor; (**d**) Nonlinear oscillatory circuit with an additional nonlinear element connected in parallel with a capacitor. On all panels: the voltage across the element N_1 nonlinearly depends on the flowing current as $V_{N1} = f(I_{N1})$, the current passing through element N_2 nonlinearly depends on the applied voltage: $I_{N2} = g(V_{N2})$

$$\begin{cases} L\dfrac{dI}{dt} = -V - f(I), \\ C\dfrac{dV}{dt} = I. \end{cases} \tag{2.1}$$

Extracting the time derivative of the left and right parts of the first equation, one can transform Eqs. (2.1) into one second-order equation:

$$\frac{d^2I}{dt^2} + \frac{1}{L}\frac{df(I)}{dI}\frac{dI}{dt} + \frac{1}{LC}I = 0. \tag{2.2}$$

If the system includes an additive resistor (Fig. 2.14b) or the wire resistance cannot be neglected, then Kirchhoff's second law becomes: $V_{N1} + V_L + V_C + V_R = 0$, which gives rise to the following system:

$$\begin{cases} L\dfrac{dI}{dt} = -V - f(I) - RI, \\ C\dfrac{dV}{dt} = I. \end{cases} \tag{2.3}$$

System (2.3) can be rewritten in the oscillatory form:

$$\frac{d^2I}{dt^2} + \frac{1}{L}\left(\frac{df(I)}{dI} + R\right)\frac{dI}{dt} + \frac{1}{LC}I = 0. \tag{2.4}$$

Nonlinearity in a RLC-circuit can be introduced by the second way. This approach consists in the connection of a nonlinear element in parallel with a capacitor (Fig. 2.14c). The current through the nonlinear element N_2 in Fig. 2.14c is assumed to be nonlinearly dependent on the applied voltage: $I_{N2} = g(V_{N2})$. The voltage drops across the capacitor C and the nonlinear element N_2 are the same, $V_C = V_{N2} = V$, since these elements are connected in parallel. Kirchhoff's first law for the upper node in Fig. 2.14c takes the form $I = I_{N2} + I_C$ or $I = g(V) + C\frac{dV}{dt}$. Kirchhoff's second law for the left loop of the circuit in Fig. 2.14c has the form $L\frac{dI}{dt} + RI + V = 0$. Finally, one can write the system of differential equations describing the circuit in Fig. 2.14c:

$$\begin{cases} L\dfrac{dI}{dt} = -V - RI, \\ \\ C\dfrac{dV}{dt} = I - g(V). \end{cases} \tag{2.5}$$

Generally, a RLC-circuit can include both elements N_1 and N_2 characterized by the dependences $V_{N1} = f(I_{N1})$ and $I_{N2} = g(V_{N2})$ (Fig. 2.14d). Then the corresponding system of differential equations becomes:

$$\begin{cases} L\dfrac{dI}{dt} = -V - f(I), \\ \\ C\dfrac{dV}{dt} = I - g(V). \end{cases} \tag{2.6}$$

The third way to introduce nonlinearity into a series oscillatory RLC-circuit is to use nonlinear feedback. This approach is discussed in Sect. 2.2.3.

2.2.2 *Nonlinear Parallel Oscillatory Circuit*

A nonlinear parallel oscillatory circuit is depicted in Fig. 2.15a. It includes a nonlinear two-terminal element N_2. The current passing through the element N_2 is assumed to be nonlinearly dependent on the voltage applied across the element: $I_{N2} = g(V_{N2})$. The voltage drops across the elements L, C, N_2 are the same $V_L = V_C = V_{N2} = V$. For the chosen directions of currents in Fig. 2.15a Kirchhoff's first law is $I_{N2} + I_L + I_C = 0$, which can be rewritten for $I = I_C$ and V in the form $C\frac{dV}{dt} + I + g(V) = 0$. Taking into account the equality $V = L\frac{dI}{dt}$, one obtains the system of two first-order differential equations:

$$\begin{cases} C\dfrac{dV}{dt} = -I - g(V), \\ \\ L\dfrac{dI}{dt} = V. \end{cases} \tag{2.7}$$

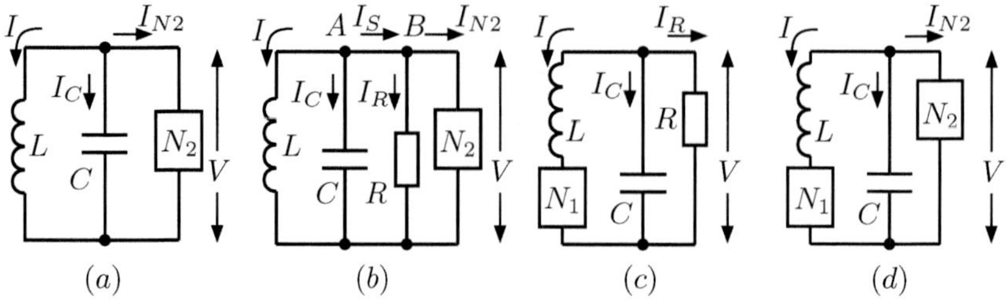

Fig. 2.15 Nonlinear parallel oscillatory circuit including (panel (**b**)) and not including (panel (**a**)) an additional resistor; (**c**) Circuit with a nonlinear element connected in series with an inductor; (**d**) Nonlinear oscillatory circuit with an additional nonlinear element connected in parallel with a capacitor. On all panels: the voltage across the element N_1 nonlinearly depends on the flowing current as $V_{N1} = f(I_{N1})$, the current passing through the element N_2 nonlinearly depends on the applied voltage: $I_{N2} = g(V_{N2})$

Extracting the time derivative of the left and right parts of the first equation, one can transform Eqs. (2.7) into one second-order equation:

$$\frac{d^2V}{dt^2} + \frac{1}{C}\frac{dg(V)}{dV}\frac{dV}{dt} + \frac{1}{LC}V = 0. \tag{2.8}$$

If the system includes an additive parallel resistor (Fig. 2.15b), then the circuit contains two nodes A and B. Kirchhoff's first law for the node A is: $I + I_C + I_S = 0$. Kirchhoff's first law for the node B becomes: $I_S = I_R + I_{N2}$ which gives rise to the equality $I_C + I + I_{N2} + I_R = 0$. Then the dynamical model equations become:

$$\begin{cases} C\dfrac{dV}{dt} = -I - g(V) - \dfrac{V}{R}, \\ L\dfrac{dI}{dt} = V. \end{cases} \tag{2.9}$$

System (2.9) can be rewritten in the oscillatory form:

$$\frac{d^2V}{dt^2} + \frac{1}{C}\left(\frac{dg(V)}{dV} + \frac{1}{R}\right)\frac{dV}{dt} + \frac{1}{LC}V = 0. \tag{2.10}$$

Nonlinearity in a parallel RLC-circuit can be introduced by the second way. One can connect a nonlinear element in series with an inductor (Fig. 2.15c). The current through the nonlinear element N_1 in Fig. 2.15c is assumed to be nonlinearly dependent on the applied voltage: $V_{N1} = f(I_{N1})$. The same current flows through the inductor L and the nonlinear element N_1, $I_L = I_{N1} = I$. Since the elements L and N_1 are connected in series, the sum of the voltage drops on these elements becomes $V_{N1} + V_L = V$ or $V = f(I) + L\frac{dI}{dt}$.

Kirchhoff's first law for the upper node of the circuit in Fig. 2.15c has the form $C\frac{dV}{dt} + \frac{V}{R} + I = 0$. Finally, one can write the system of differential equations describing the circuit in Fig. 2.15c:

$$\begin{cases} C\dfrac{dV}{dt} = -I - \dfrac{V}{R}, \\ L\dfrac{dI}{dt} = V - f(I). \end{cases} \tag{2.11}$$

Generally, a RLC-circuit can include both elements N_1 and N_2 characterized by the dependences $V_{N1} = f(I_{N1})$ and $I_{N2} = g(V_{N2})$ (Fig. 2.15d). Then the corresponding system of differential equations becomes:

$$\begin{cases} C\dfrac{dV}{dt} = -I - g(V), \\ L\dfrac{dI}{dt} = V - f(I). \end{cases} \tag{2.12}$$

Similarly to a series oscillatory circuit, the third way to introduce nonlinearity into a parallel oscillatory RLC-circuit is to use nonlinear feedback. This approach is discussed in Sect. 2.2.4.

Equations describing the series and parallel oscillatory RLC-circuits are fully equivalent to each other. Choice of either configuration depends on the specifics of nonlinear elements and measurements.

2.2.3 Series Oscillatory Circuit with Feedback

The circuit depicted in Fig. 2.16a is a non-autonomous series RLC-circuit including a source of an electromotive force ε. One can distinguish cases $\varepsilon = \text{const}$ (fixed value), $\varepsilon = \varepsilon(t)$ (time-dependent signal). What if the electromotive force value depends on an instantaneous value of a dynamical variable, for instance on a voltage across a capacitor or on a current through an inductor? This particular case can be realized using feedback.

The circuit shown in Fig. 2.16b includes feedback organized by a nonlinear amplifier. The amplifier output voltage depends on the voltage across the capacitor: $V_f = f(V)$. The amplifier output voltage is fed back into the RLC-loop. The amplifier input resistance is assumed to be extremely large. Thus, the amplifier input current I_{in} tends to zero, $I_{\text{in}} = 0$. Then the same current I passes through the resistor R, the capacitor C, the inductor L. Finally, Kirchhoff's second law for the RLC-loop becomes: $L\frac{dI}{dt} + IR + V = V_f(V)$. The corresponding system of differential equations for dynamical variables I and V is written as:

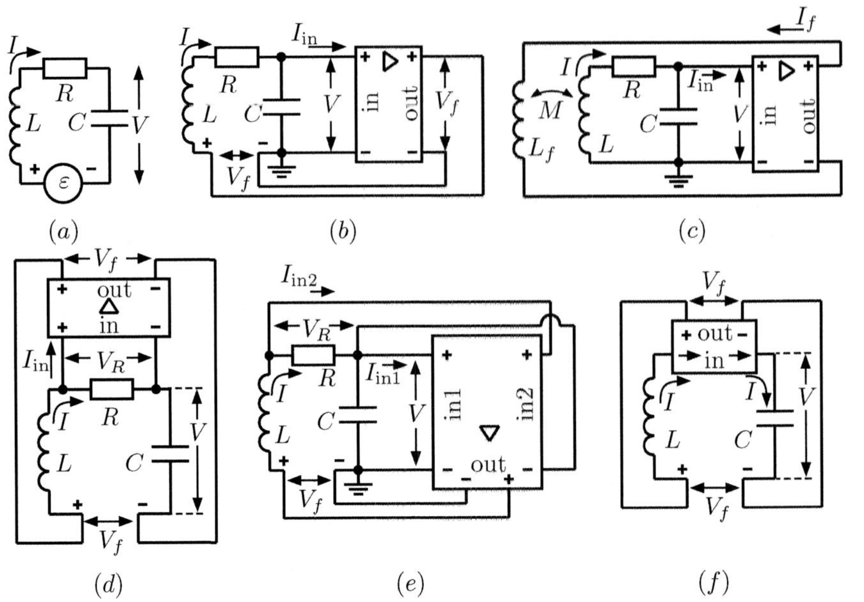

Fig. 2.16 Series RLC-circuit: non-autonomous circuit with a source of an electromotive force ε (panel (**a**)) and driven by different feedbacks (panels (**b**)—(**f**))

$$\begin{cases} L\dfrac{dI}{dt} + IR + V = f(V), \\ C\dfrac{dV}{dt} = I, \end{cases} \tag{2.13}$$

which can be transformed into the oscillatory form:

$$\frac{d^2V}{dt^2} + \frac{R}{L}\frac{dV}{dt} + \frac{1}{LC}V = \frac{1}{LC}f(V). \tag{2.14}$$

The feedback loop involving the variable V can be organized in different ways. One of the possible realizations is depicted in Fig. 2.16c. Here the amplifier output signal is the current $I_f = f(V)$ flowing through the inductor L_f. As in the previous case, the amplifier input current is neglected, $I_{\text{in}} = 0$. The inductors L and L_f are situated close to each other such that changes of the magnetic field in the inductor L_f produces the potential difference ε across the inductor L (thus, the circuit in Fig. 2.16c contains a mutual inductance M). In the simplest case the potential difference ε is assumed to be linearly proportional to the first time derivative of the amplifier output current: $\varepsilon = M\frac{dI_f(V)}{dt} = M\frac{df(V)}{dV}\frac{dV}{dt}$, where M is a coefficient of inductive coupling. Then Kirchhoff's second law for the RLC-loop becomes:

$L\frac{dI}{dt} + IR + V = M\frac{df}{dV}\frac{dV}{dt}$. Taking into account the formula $I = C\frac{dV}{dt}$, one obtains the equation:

$$\frac{d^2V}{dt^2} + \left(\frac{R}{L} - \frac{M}{LC}\frac{df(V)}{dV}\right)\frac{dV}{dt} + \frac{1}{LC}V = 0. \tag{2.15}$$

One can realize feedback involving the nonlinear transformation of an instantaneous value of the current I. For this purpose the voltage drop across the resistor R is used as an amplifier input signal (Fig. 2.16d). The amplifier input current is neglected as before, $I_{\text{in}} = 0$. Then the output voltage V_f fed back into the RLC-loop nonlinearly depends on the voltage across the resistor R as being: $V_f = V_f(V_R) = V_f(IR)$. Since the resistance R is assumed to be constant, one can write: $V_f = f(I)$. Then Kirchhoff's second law for the circuit becomes: $L\frac{dI}{dt} + IR + V = f(I)$. Extracting the time derivative, one derives the equality $L\frac{d^2I}{dt^2} + R\frac{dI}{dt} + \frac{dV}{dt} = \frac{df(I)}{dI}\frac{dI}{dt}$. Using the formula $I = C\frac{dV}{dt}$, one obtains the final equation describing the system:

$$\frac{d^2I}{dt^2} + \frac{1}{L}\left(R - \frac{df(I)}{dI}\right)\frac{dI}{dt} + \frac{1}{LC}I = 0. \tag{2.16}$$

The feedback loop can takes more complicated forms. For example, the amplifier output voltage can be determined by both dynamical variables V and I. Such situation involving both feedback approaches discussed above is depicted in (Fig. 2.16e), where the amplifier output voltage is a function $V_f = f(V, I)$. The input currents are neglected, $I_{\text{in}1} = I_{\text{in}2} = 0$. Then Kirchhoff's second law for the circuit is $L\frac{dI}{dt} + IR + V = f(V, I)$ and the final system of differential equations becomes:

$$\begin{cases} L\dfrac{dI}{dt} = f(V, I) - IR - V, \\[2ex] C\dfrac{dV}{dt} = I. \end{cases} \tag{2.17}$$

Further steps for the transformation into the oscillatory form depend on a concrete form of the function $f(V, I)$.

It is not necessary to use the voltage V_R for the creation of feedback involving dependences on the current I. In some cases it is problematic to add a resistor, because it can change the properties of the system under study. There exist other circuits of current-voltage transformers which do not amend the resistive properties of explored circuits. In such a case an oscillatory circuit driven by feedback can be simplified, for instance as in Fig. 2.16f. The equation describing the circuit is:

$$\frac{d^2I}{dt^2} - \frac{1}{L}\frac{df(I)}{dI}\frac{dI}{dt} + \frac{1}{LC}I = 0. \tag{2.18}$$

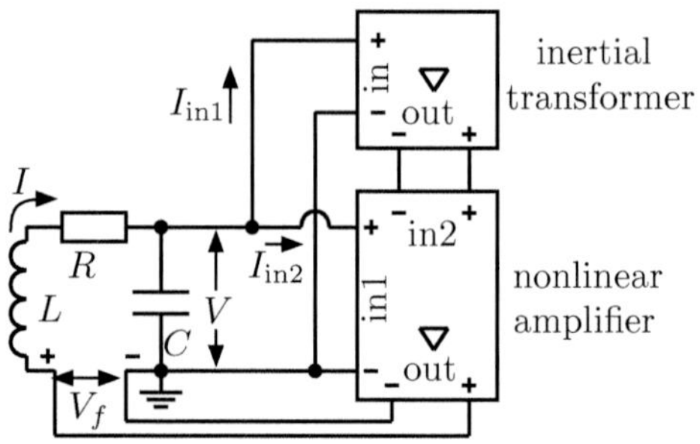

Fig. 2.17 Series RLC-circuit with inertial nonlinear feedback

The introduction of complex feedback can lead to the appearance of new dynamical variables. One of such configurations is depicted in Fig. 2.17. In this circuit the voltage V across the capacitor is used as the input signal of two blocks: the inertial transformer and the nonlinear amplifier. The corresponding input currents are assumed to be zero, $I_{\text{in}1,2} = 0$. The inertial transformer produces the output voltage V_1 such that $V_1 = \int_0^t f(V)dt$ or $\frac{dV_1}{dt} = f(V)$. The signal V_1 is used as the additional input of the nonlinear amplifier. The nonlinear amplifier output voltage V_f depends on the instantaneous values of V and V_1 such that $V_f = g(V, V_1)$. Using Kirchhoff's second law for the RLC-loop written in the form $L\frac{dI}{dt} + IR + V = g(V, V_1)$, one can derive the system of three first-order differential equations for three dynamical variables I, V, V_1:

$$\begin{cases} L\dfrac{dI}{dt} = g(V, V_1) - IR - V, \\ C\dfrac{dV}{dt} = I, \\ \dfrac{dV_1}{dt} = f(V). \end{cases} \tag{2.19}$$

This approach is discussed in paper [6] in the context of Anishchenko-Astakhov chaotic self-oscillator with inertial nonlinearity.

2.2.4 *Parallel Oscillatory Circuit with Feedback*

Figure 2.18 illustrates different options for the introduction of nonlinear feedback in a parallel RLC-circuit. The circuit in Fig 2.18a contains the amplifier the input signal of which is the voltage across the capacitor, V. The amplifier output voltage nonlinearly depends on the input voltage $V_f = f(V)$. The output signal returns into the RCL-circuit by using the resistor R_f. Then the voltage drop across the

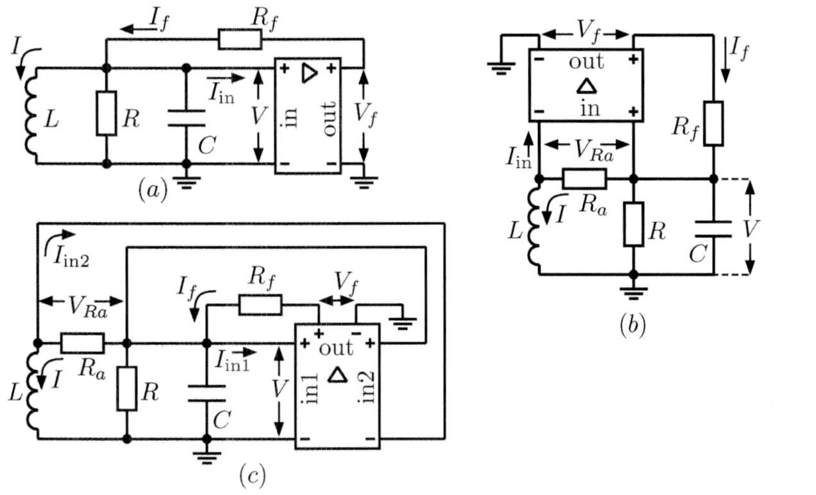

Fig. 2.18 Different options for the introduction of nonlinear feedback into a parallel RLC-circuit

resistor R_f becomes $\Delta V_{Rf} = V_f - V$. It induces the current through the resistor R_f being equal to $I_f = (V_f - V)/R_f$. The amplifier input current I_{in} is assumed to be extremely small. Then Kirchhoff's first law for one of upper nodes becomes: $I + \frac{V}{R} + C\frac{dV}{dt} = I_f$. Using the equality $L\frac{dI}{dt} = V$ and extracting the time derivative, one can rewrite Kirchhoff's first law expression into the form:

$$\frac{d^2V}{dt^2} + \left(\frac{1}{RC} + \frac{1}{R_f C}\left(1 - \frac{df}{dV}\right)\right)\frac{dV}{dt} + \frac{1}{LC}V = 0. \tag{2.20}$$

As in the case of a series RLC-circuit, feedback can involve the current through the inductor of a parallel oscillatory circuit. For this purpose one can use a special current-voltage transformer or add a resistor in series with the inductor as in Fig. 2.18b. In such a case the amplifier output voltage signal is $V_f = V_f(V_{Ra}) = V_f(R_a I) = f(I)$. As before, we assume $I_{\text{in}} = 0$. Using Kirchhoff's first law, one can derive the circuit equation for the current through the inductor, I, and the voltage across the capacitor, V:

$$\begin{cases} C\dfrac{dV}{dt} = -\dfrac{V}{R} - I + \dfrac{f(I) - V}{R_f}, \\[2ex] L\dfrac{dI}{dt} = V - IR_a. \end{cases} \tag{2.21}$$

If the feedback nonlinear transformation involves both variables, the voltage V and the current I (Fig. 2.18c), then the amplifier output voltage is a function $V_f = f(I, V)$ and the corresponding equations are:

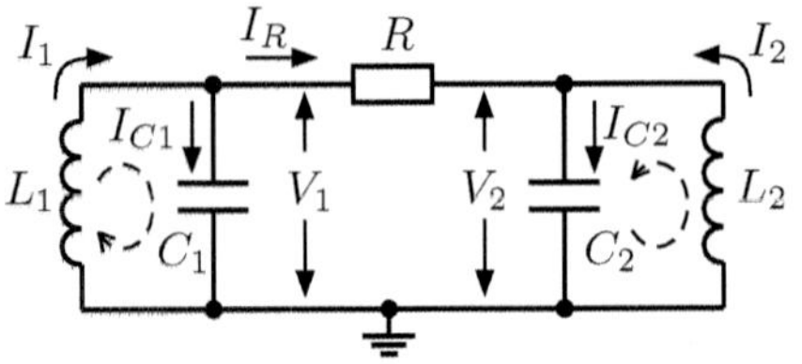

Fig. 2.19 Exemplary circuit of two coupled LC-loops.

$$\begin{cases} C\dfrac{dV}{dt} = -\dfrac{V}{R} - I + \dfrac{f(I,V) - V}{R_f}, \\ L\dfrac{dI}{dt} = V - IR_a. \end{cases} \tag{2.22}$$

2.2.5 *Coupled RLC-Circuits*

If a circuit contains several oscillatory loops, then it is described by a multidimensional model including several voltages and currents as dynamical variables. The loops can be coupled by different ways. One exemplary configuration is depicted in Fig. 2.19. Here, two linear LC-circuits are resistively coupled. For chosen directions of all currents and traversals along each circuit loop Kirchhoff's first and second laws take the following forms:

$$I_1 = I_{C1} + I_R, \quad I_2 + I_R = I_{C2}, \quad L_1\frac{dI_1}{dt} + V_1 = 0, \quad L_2\frac{dI_2}{dt} + V_2 = 0, \tag{2.23}$$

which are accompanied by the equalities: $I_{C1} = C_1\frac{dV_1}{dt}$, $I_{C2} = C_2\frac{dV_2}{dt}$, $(V_1 - V_2)/R = I_R$. Then Eqs. (2.23) are transformed into the form:

$$\begin{cases} I_1 = C_1\dfrac{dV_1}{dt} + \dfrac{V_1 - V_2}{R}, \\ I_2 + \dfrac{V_1 - V_2}{R} = C_2\dfrac{dV_2}{dt}, \\ L_1\dfrac{dI_1}{dt} + V_1 = 0, \\ L_2\dfrac{dI_2}{dt} + V_2 = 0, \end{cases} \tag{2.24}$$

Dynamical model (2.24) is fully defined: it includes four differential equations with four dynamical variables $I_{1,2}$ and $V_{1,2}$.

2.3 Wien Bridge Oscillator

An electronic oscillator based on Wien bridge is schematically depicted in Fig. 2.20. It contains a nonlinear amplifier and a Wien bridge circuit which provides positive feedback. The amplifier output voltage is a function of the input voltage: $V_f = f(V)$. The voltage drops across the elements R_1 and C_1 constitute the difference between the voltages V_f and V which is written as $V = V_f - V_{R1} - V_{C1}$. The same current I flows through the elements R_1 and C_1. Then the corresponding voltage drops are $V_{R1} = IR_1$ and $V_{C1} = \frac{1}{C_1}\int_0^t I dt$. The amplifier input current is neglected. Kirchhoff's first law allows to write the equality for currents: $I = I_{R2} + I_{C_2}$ which can be rewritten in the form $I = \frac{V}{R_2} + C_2\frac{dV}{dt}$. The equalities for voltages and currents are combined:

$$V = V_f - IR_1 - \frac{1}{C_1}\int_0^t I dt, \quad I = \frac{V}{R_2} + C_2\frac{dV}{dt}. \tag{2.25}$$

Substituting the second equality into the first one and extracting the time derivative, one obtains the equation describing the circuit in terms of the voltage V:

$$\frac{d^2V}{dt^2} + \left(\frac{1}{R_1C_2} + \frac{1}{R_2C_2} + \frac{1}{R_1C_1} - \frac{1}{R_1C_2}\frac{df(V)}{dt}\right)\frac{dV}{dt} + \frac{1}{C_1C_2R_1R_2}V = 0. \tag{2.26}$$

In particular case, $C_1 = C_2 = C$, $R_1 = R_2 = R$, Eq. (2.26) is simplified:

$$\frac{d^2V}{dt^2} + \frac{1}{RC}\left(3 - \frac{df(V)}{dt}\right)\frac{dV}{dt} + \frac{1}{C^2R^2}V = 0. \tag{2.27}$$

If the amplifier is linear, $f(V) = KV$, then Eq. (2.27) describes a linear oscillator. In case $K < 3$ model (2.27) exhibits periodic damped oscillations. If the gain is greater than three, $K > 3$, then self-oscillation excitation occurs. The self-oscillation

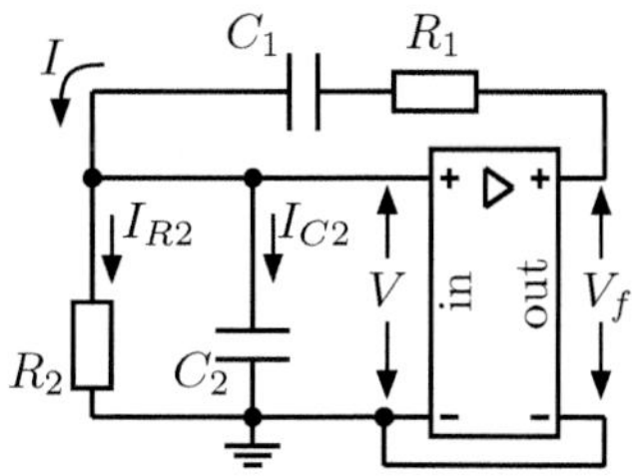

Fig. 2.20 Wien bridge oscillator

frequency equals to $\omega_0 = (RC)^{-1}$. The oscillation amplitude increases while limitation mechanisms (nonlinearity or the amplifier output voltage limitation) do not act.

2.4 External Force and Dimensionality Increase

The circuit in Fig. 2.17 demonstrates how to increase the dimensionality of a RLC-circuit mathematical model using the feedback properties. Multidimensional oscillatory circuits based on the RLC-loop can be implemented in a different manner. For this purpose one can use external influences. Let us reconsider the circuit in Fig. 2.16a. The circuit equations are $L\frac{dI}{dt} + RI + V = \varepsilon$ and $C\frac{dV}{dt} = I$. From this point of view the dynamical system is described by two dynamical variables I and V. However the instantaneous value of the external forcing, $\varepsilon(t)$, can vary according to some dynamical law. Then the system dimensionality can increase. For example, the sinusoidal signal $\varepsilon = \varepsilon_{\max}\sin(\omega t)$ can be rewritten as $\varepsilon = \varepsilon_{\max}\sin(z)$, where $z = \omega t$ is considered as a new dynamical variable determined by the equation $\frac{dz}{dt} = \omega$. In such a case the model equations are:

$$\begin{cases} L\dfrac{dI}{dt} = -RI - V + \varepsilon_{\max}\sin(z), \\ C\dfrac{dV}{dt} - I, \\ \dfrac{dz}{dt} = \omega. \end{cases} \tag{2.28}$$

External signals can be stochastic. In such cases dimensionality growth is also possible. For example, the action of a coloured noise source can be described by the first-order Ornstein-Uhlenbeck process (see Sect. 4.2.2 where the issue of the Ornstein-Uhlenbeck process electronic implementation is briefly discussed): $\tau_c\frac{d\varepsilon}{dt} = -\varepsilon + \sqrt{2D\tau_c}n(t)$, where $n(t)$ is a source of white Gaussian noise with the intensity D, and τ_c is the coloured noise correlation time. Then the stochastically driven RLC-circuit in Fig. 2.16a is described by the following three-dimensional model:

$$\begin{cases} L\dfrac{dI}{dt} = -RI - V + \varepsilon, \\ C\dfrac{dV}{dt} = I, \\ \tau_c\dfrac{d\varepsilon}{dt} = -\varepsilon + \sqrt{2D\tau_c}n(t). \end{cases} \tag{2.29}$$

The same methodology can be used in the context of different kinds of circuits such like nonlinear RLC-circuits, coupled RLC-circuits or Wien bridge oscillator in order to increase the system dimensionality and to induce more complicated dynamical regimes.

References

1. M. Jones, *A Practical Introduction to Electronic Circuits*, 3rd edn. (Cambridge University Press, Cambridge, 1995)
2. P. Horowitz, W. Hill, *The Art of Electronics*, 3rd edn. (Cambridge University Press, Cambridge, 2015)
3. R. Boylestad, L. Nashelsky, *Electronic Devices and Circuit Theory*, 8th edn. (Prentice Hall, Saddle River, 2001)
4. R. Tetzlaff (ed), *Memristors and Memristive Systems* (Springer, New York, 2014)
5. S. Vaidyanathan, C. Volos, *Advances in Memristors, Memristive Devices and Systems* (Springer, Berlin, 2017)
6. I. Korneev, I. Ramazanov, A. Slepnev, T. Vadivasova, V. Semenov, Chaos **33**(9), 091102 (2023)

Chapter 3
Analog Modelling and Signal Processing

Many of the circuits considered in the second chapter contain nonlinear converters and amplifiers mentioned as 'black boxes' with certain characteristics and output functions. It is explained in this chapter how to implement these boxes by means of analog circuits. After that the second approach of electronic modelling of dynamical systems, **analog modelling**, is described. Finally, methods of the implementation of two-terminal circuits with specified characteristics are presented. Before the study of analog circuits begins, used integrated circuits and their functioning aspects are discussed. The chapter is mostly focused on materials related to electronic modelling issues. Books [1–3] are recommended for additional study and knowledge extension.

3.1 Analog Integrated Circuits

3.1.1 Operational Amplifier

An operational amplifier (op-amp) is a DC-coupled differential, single-ended, high gain amplifier. The op-amps are integrated circuits composed of many elements (transistors, resistors, etc.) such that the resulting circuit follows certain laws. The op-amp produces an output potential (relative to the circuit ground) being proportional to the potential difference between its input terminals. The simplest depiction of the op-amp is shown in Fig. 3.1a. The depiction with power supply terminals (Fig. 3.1b) is also possible. Schematically, op-amp functioning can be illustrated as in Fig. 3.1c. Two inputs marked by symbols '-' and '+' are called 'inverting' and 'non-inverting'. For the voltages V_{in1} and V_{in2} applied to the non-inverting and inverting inputs correspondingly (Fig. 3.1) the amplifier produces the output voltage $V_{\text{out}} = G(V_{\text{in1}} - V_{\text{in2}}) + V_{\text{offset}}$, where G is a very high open-loop gain coefficient, V_{offset} is an offset voltage generated by the amplifier. Amplifier's input

© The Author(s), under exclusive license to Springer Nature Switzerland AG 2024

V. Semenov, *Electronic Modelling of Deterministic and Stochastic Oscillators*,
Springer Series in Synergetics, https://doi.org/10.1007/978-3-031-81553-9_3

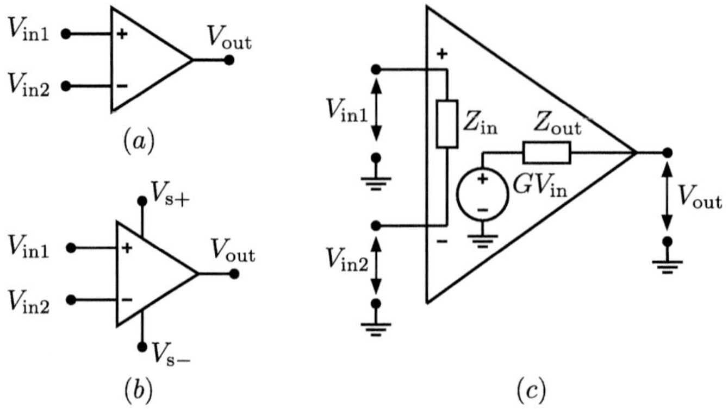

Fig. 3.1 Operational amplifier depiction without (panel (**a**)) and with (panel (**b**)) power supply pins; (**c**) Schematically represented op-amp functioning

impedance, Z_{in}, is assumed to be very high. Consequently, the input currents are extremely small. The output impedance, Z_{out}, is assumed to be very small which allows to support high the output current. The ideal op-amp model implies the following assumptions:

- Infinite open-loop gain, $G = \infty$;
- Infinite input impedance, $Z_{\text{in}} = \infty$. As a result, the input currents are zero, $I_{\text{in}} = 0$;
- Zero output impedance, $Z_{\text{out}} = 0$. As a result, the output current range is not limited;
- Zero offset voltage, $V_{\text{offset}} = 0$;
- Infinite output voltage range;
- Infinite bandwidth where output signal characteristics do not depend on the input signal frequency.

Real op-amps are not ideal and are characterized by the following properties:

- Very high but finite open-loop gain. Typical values are $G \in [10^5, 10^6]$;
- Very high but finite input impedance, up to several megohms and above. As a result, the input currents persist but are small (from a few picoamps to a few milliamps);
- Small, but nonzero output impedance. The amplifier does not correctly work when the output current increases and passes through maximal allowable value;
- Nonzero offset voltage, up to millivolts;
- Finite output voltage range limited by the voltage supply, $V_{\text{out}} \in [V_{\text{s}-}, V_{\text{s}+}]$;
- Finite bandwidth where output signal characteristics can be assumed to be indepent on the input signal frequency. Outside from an allowable frequency range one deals with phase shifts, signal distortions and imperfections associated with amplifier's intrinsic peculiarities and finite slew rate.

The op-amp was intended for use with external feedback elements, which determine the resultant function (or operation). When negative feedback is used, circuit's output response is determined mostly by the feedback network, rather than by the op-amp characteristics. In the presence of the negative feedback the op-amp gets a very important property: the op-amp controls the input terminal states and tends to eliminate the voltage difference between the inverting and non-inverting inputs. In other words, the op-amp creates the output voltage such that the resulting input potential difference (voltage difference $V_{\text{in}1} - V_{\text{in}2}$ in Fig. 3.1) tends to zero. This rule will be used in Sect. 3.2 for the derivation of equations describing particular configurations of the op-amp driven by negative feedback.

3.1.2 Analog Comparator

Generally, a comparator is a device that compares two input voltage (or current is some special cases) values and generates a two-level output signal indicating which input voltage is larger. An operational amplifier without feedback can be used for the comparator function implementation. Indeed, if the op-amp depicted in Fig. 3.1 has two input voltages $V_{\text{in}1}$ and $V_{\text{in}2}$. The output voltage is $V_{\text{out}} = G(V_{\text{in}1} - V_{\text{in}2})$. Since the gain coefficient G is very high, the output voltage also possesses very high values. However, the output voltage range of real op-amps is limited by a source voltage value. Thus, op-amp's output signal takes two values: $V_{\text{s}+}$ for $V_{\text{in}1} - V_{\text{in}2} > 0$ (which corresponds to $V_{\text{in}1} > V_{\text{in}2}$) and $V_{\text{s}-}$ for $V_{\text{in}1} - V_{\text{in}2} < 0$ (which corresponds to $V_{\text{in}1} < V_{\text{in}2}$). If the value $V_{\text{in}1}$ is very close $V_{\text{in}2}$, then one can observe spontaneous transitions between $V_{\text{s}-}$ and $V_{\text{s}-}$ caused by noise and the comparator operation specifics.

The comparator action is illustrated in Fig. 3.2a. The depicted circuit has two inputs: $V_{\text{in}1} = V_{\text{in}} = A\cos(\omega t)$ (the non-inverting input is connected to an AC-voltage source) and $V_{\text{in}2} = 0$ (the inverting input is connected to ground). The output voltage is $V_{\text{out}} = V_{\text{s}+}$ for $V_{\text{in}1}(t) > 0$ and $V_{\text{out}} = V_{\text{s}-}$ for $V_{\text{in}1}(t) < 0$ (Fig. 3.2b). Transient time corresponding to transition between two states $V_{\text{s}-}$ and $V_{\text{s}+}$ depends on the op-amp characteristics. Particularly, the op-amp action speed

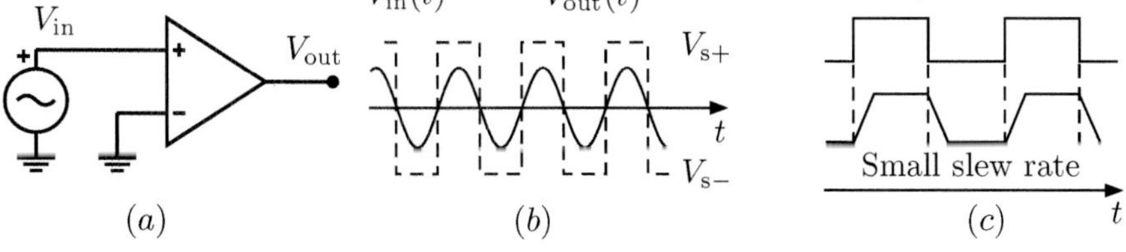

Fig. 3.2 (**a**) Comparator with an AC-voltage input; (**b**) Time realizations of the comparator input (solid line) and output (dashed line) voltages; (**c**) Response of a comparator with large (upper realization) and small (lower realization) slew rate

is specified by a quantity 'slew rate'. The op-amp characteristics 'slew rate' is defined as the maximal possible change of voltage per unit of time. The 'slew rate' is usually expressed in $[V/\mu s]$. If the op-amp slew rate is insufficient, then transients become visible (Fig. 3.2c). To decrease transient time, one can use high-speed op-amps (for example, the op-amp AD826AN) or specially produced integrated circuit comparator (for example, the comparator AD790JNZ).

3.1.3 Analog Multiplier

Different circuits implementing the multiplication of two analog signals are considered in Sect. 3.2.11. Here we focus on integrated circuits realizing the multiplication. If two voltage signals $V_1(t)$ and $V_2(t)$ are connected to multiplier input ports, the output voltage contains the product of multiplication $V_1 \times V_2$ divided by a scaling factor. The scaling factor dimension is the voltage. Usually, the multiplier input impedance is assumed to be infinitely large while the output impedance is assumed to be zero. Depending on an input voltage range where the multiplier works correctly, one distinguishes three types of multipliers: single quadrant (V_1 and V_2 are unipolar), two-quadrant (V_1 is unipolar, V_2 is bipolar), four-quadrant (V_1 and V_2 are bipolar).

Further study of the analog multiplier action is carried out on the example of the integrated circuit AD633JN. The device has eight terminals (Fig. 3.3a): five input terminals (four terminals for input voltages $X1$, $X2$, $Y1$ and $Y2$ and summing input Z), one output terminal (output voltage V_{out}) and two power supply terminals (power supply voltages can range from $\pm$ 8 to $\pm$ 18 V). The output voltage value $V_{\text{out}} = \frac{(V_1-V_2)(V_3-V_4)}{10} + Z$. The scale factor 10 [V] inhibits the large output voltage appearance. The summing input has two functions. On the one hand, it allows to use an additive signals. On the other hand, it makes be possible to organize output feedback. Particularly, the feedback in Fig. 3.3b represents a voltage divider which takes the output voltage V_{out} and sends the fraction $\frac{R_2}{R_1+R_2}V_{\text{out}}$ to the summing input Z. Then the output voltage formula becomes $V_{\text{out}} = \frac{(V_1-V_2)(V_3-V_4)}{10} + \frac{R_2}{R_1+R_2}V_{\text{out}}$

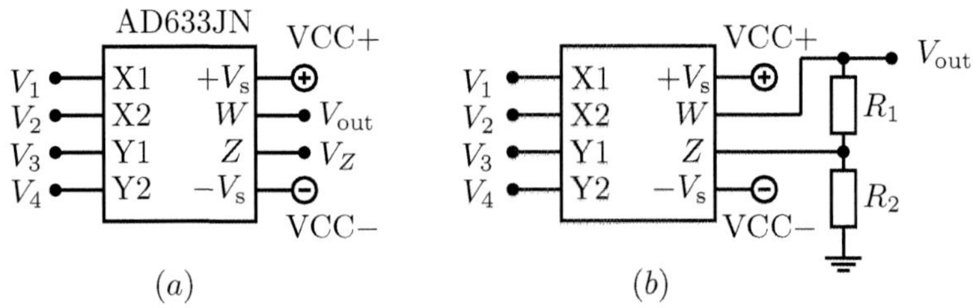

Fig. 3.3 (**a**) Analog multiplier AD633JN pin configuration; (**b**) AD633JN multiplier with an output feedback loop

which can be rewritten in the form $V_{out} = \frac{(V_1-V_2)(V_3-V_4)}{10}\frac{R_1+R_2}{R_1}$. If, for example, $R_1 = 1\,\text{k}\Omega$ and $R_2 = 9\,\text{k}\Omega$, then $V_{out} = (V_1 - V_2)(V_3 - V_4)$. As you can see, the feedback allows to amplify the output voltage without external amplifiers. It is important to note that in addition to the multiplier functions, AD633JN is widely used for the analog circuit implementation of division, amplitude modulation, frequency doubling, squaring and square rooting, and many other procedures (see the datasheet for AD633JN).

3.2 Electronic Implementation of Mathematical Operations

3.2.1 Voltage Follower (Unity-Gain Amplifier)

The operational-amplifier-based voltage follower (also called a unity-gain amplifier, a buffer amplifier, and an isolation amplifier) is depicted in Fig. 3.4a. Here, the output voltage directly follows the input one. In the context of the voltage signal processing, the voltage follower does nothing. The purpose of a voltage follower is hidden in the input-output currents. The voltage follower has a very high input impedance (input currents are extremely low). Then a very little current is drawn from any circuit and its impact on the circuit is reduced (for instance, see the circuit for the voltage signal acquisition in Fig. 3.4b). In contrast, follower's output impedance is very small which provides for correct connection to measuring and communication equipment (for instance, to analog oscilloscope in Fig. 3.4b).

Generally, the voltage follower is useful and is connected between a circuit and a load when the low impedance load demands and draws a high current. In the absence of the voltage follower the low impedance load consumes huge amounts of power to be drawn from the circuit as from a power source and, because of this, causes high disturbances and incorrect use of the circuit.

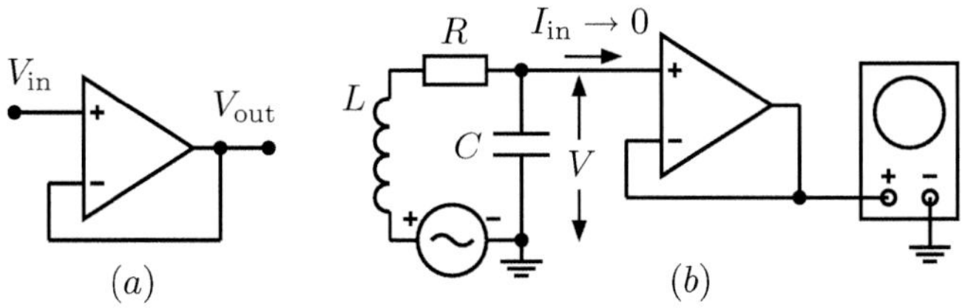

Fig. 3.4 Voltage follower based on an operational amplifier (panel (**a**)) and its usage for the voltage signal acquisition (panel (**b**))

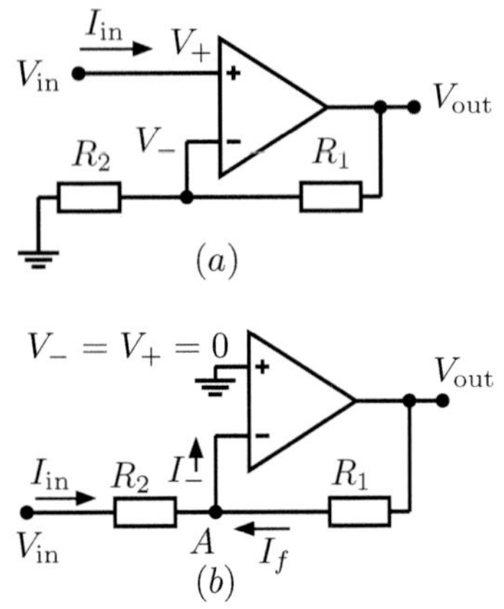

Fig. 3.5 Non-inverting (panel (**a**)) and inverting (panel (**b**)) amplifier based on an op-amp

3.2.2 Multiplication by a Constant (Inverting and Non-inverting Amplifier)

An op-amp-based non-inverting amplifier circuit is depicted in Fig. 3.5a. Since an op-amp tends to eliminate the voltage difference between the inverting and non-inverting inputs, one can assume $V_- = V_+ = V_{in}$. The feedback loop containing the resistors R_1 and R_2 is a voltage divider which is described by the formula $V_+ = V_{in} = \frac{R_2}{R_1+R_2} V_{out}$. Then the output voltage is $V_{out} = \left(1 + \frac{R_1}{R_2}\right) V_{in}$. It is important to repeat the principle: the op-amp connected with negative feedback produces the output voltage such that $V_- - V_+ = 0$. This condition is satisfied when $V_{out} = \left(1 + \frac{R_1}{R_2}\right) V_{in}$. The circuit input impedance which determines the input current I_{in} equals to the op-amp input impedance given in a datasheet. Circuit's gain value $G = 1 + \frac{R_1}{R_2} > 1$ is always positive. For this reason the circuit in Fig. 3.5a is called a 'non-inverting amplifier'.

An op-amp-based inverting amplifier (also known simply as an inverter) circuit is depicted in Fig. 3.5b. As in the previous circuit, the condition takes place $V_- = V_+$ takes place. Since the non-inverting input is connected to ground, the condition is complemented $V_- = V_+ = 0$. Then the current I_{in} flowing through the resistor R_2 becomes: $I_{in} = (V_{in} - V_-)/R_2 = V_{in}/R_2$. Similarly, the current I_f passing through the resistor R_1 equals to $(V_{out} - V_-)/R_1 = V_{out}/R_1$. Kirchhoff's first law for the node A in Fig. 3.5b takes the form: $I_{in} + I_f = I_-$ or $I_{in} = -I_f$ (because the op-amp input current is supposed to be zero). Substituting formulas for currents, one obtains: $V_{out}/R_1 = -V_{in}/R_2$ or $V_{out} = -\frac{R_2}{R_1} V_{in}$. The sign '−' means that the amplification of the input signal is accompanied by inverting. For this reason the circuit in Fig. 3.5b is called an 'inverting amplifier'. The input impedance of the circuit in Fig. 3.5b equals to the resistance R_2. If the high input current problem is critical, one can add the voltage follower described in Sect. 3.2.1 before the resistor

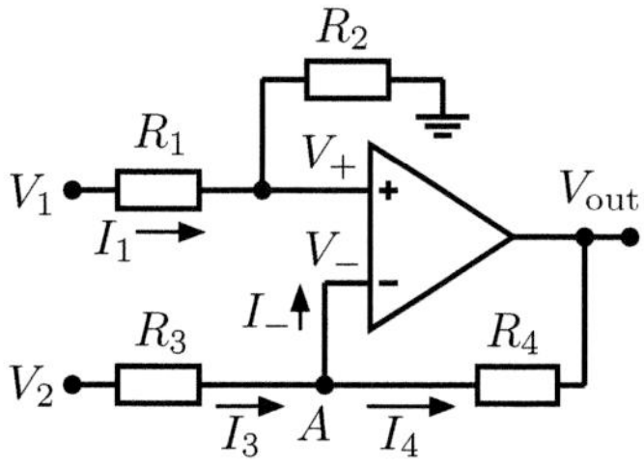

Fig. 3.6 Op-amp-based differential amplifier

R_2. Then the circuit input impedance is sufficiently increased which results in strong decrease of the circuit input current.

3.2.3 Subtraction (Differential Amplifier)

First of all, one should not confuse a 'differential amplifier' and an 'op-amp-based differentiator' considered in Sect. 3.2.8. The differential amplifier output signal contains the difference between two input voltage signals. In comparison to a single op-amp, the gain factor of the differential amplifier is much smaller, and the op-amp saturation is not reached. The op-amp-based differential amplifier circuit is depicted in Fig. 3.6. The resistors R_1 and R_2 make up a voltage divider and the voltage at the op-amp non-inverting input is $V_+ = \frac{R_2}{R_1+R_2}V_1$. The voltage at the op-amp inverting input is the same $V_- = V_+ = \frac{R_2}{R_1+R_2}V_1$ since the op-amp is connected to negative feedback. Then the current flowing through the resistor R_3 becomes: $I_3 = \frac{V_2-V_-}{R_3} = \frac{V_2}{R_3} - \frac{R_2V_1}{R_3(R_1+R_2)}$. The current passing through the resistor R_4 is $I_4 = \frac{V_- - V_{out}}{R_4} = \frac{R_2V_1}{R_4(R_1+R_2)} - \frac{V_{out}}{R_4}$. Kirchhoff's first law for the node A in Fig. 3.6 takes the form $I_3 = I_4$ because the op-amp input current is assumed to be zero. Then one can substitute formulas for the currents $I_{3,4}$ and obtain the equality $\frac{V_2}{R_3} - \frac{R_2V_1}{(R_1+R_2)R_3} = -\frac{V_{out}}{R_4} + \frac{R_2V_1}{(R_1+R_2)R_4}$. After that the output voltage can be expressed: $V_{out} = \frac{R_2(R_3+R_4)}{R_3(R_1+R_2)}V_1 - \frac{R_4}{R_3}V_2$. In particular case $R_1 = R_3 = R_{01}$, $R_2 = R_4 = R_{02}$ the output voltage formula is simplified: $V_{out} = \frac{R_{02}}{R_{01}}(V_1 - V_2)$.

Since the op-amp input currents are neglected, the same current I_1 passes through the resistors R_1 and R_2. The total voltage drop across these resistors equals to V_1. Then the circuit input current I_1 is $V_1/(R_1 + R_2)$ and the corresponding input impedance becomes $Z_1 = \frac{V_1}{I_1} = R_1 + R_2$. Similarly, the same current flows through the resistors R_4 and R_4. Then $I_3 = (V_2 - V_{out})/(R_3 + R_4)$ and the input impedance becomes $Z_2 = (V_2 - V_{out})/I_3 = R_3 + R_4$. As you can see, the input impedances are finite. This disadvantage can be avoided by adding the voltage followers described above before the resistors R_1 and R_3. The second way is changing the differential amplifier circuit discussed in this section to the instrumentation amplifier circuit.

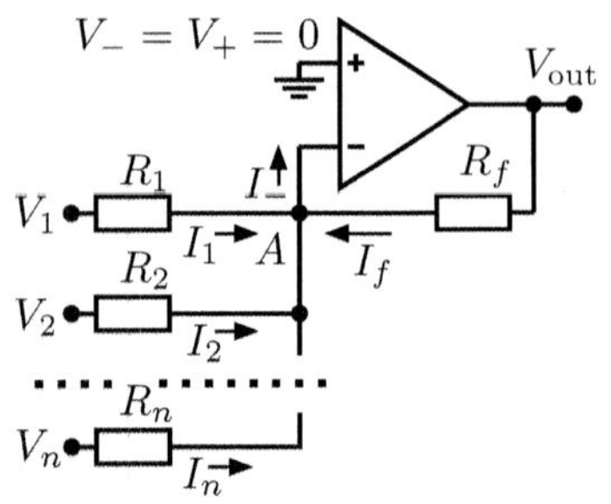

Fig. 3.7 Op-amp-based summing amplifier

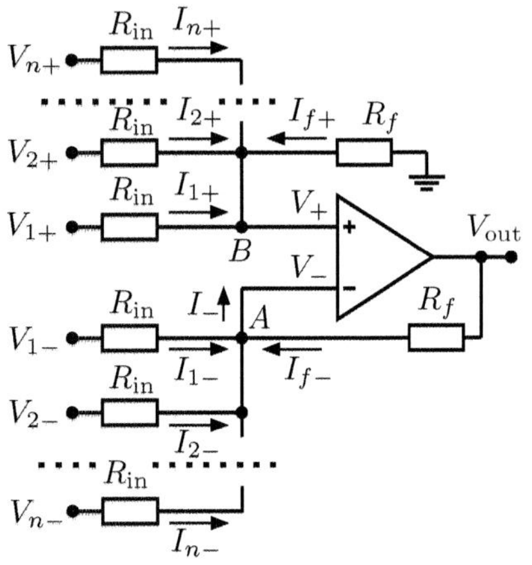

Fig. 3.8 Circuit implementation of simultaneous addition and subtraction

3.2.4 *Addition (Summing Amplifier)*

An op-amp-based summing amplifier circuit is depicted in Fig. 3.7. The circuit is similar to an op-amp-based inverting amplifier (see Fig. 3.5b). The circuit contains several resistors R_1, R_2,…,R_n making up the circuit inputs. The input voltages are V_1, V_2,…,V_n. Kirchhoff's first law for the node A in Fig. 3.7 takes the form: $I_1 + I_2 + \ldots + I_n + I_f = I_-$ or $I_1 + I_2 + \ldots + I_n = -I_f$ (because the op-amp input current is supposed to be zero). Substituting formulas for currents, one obtains: $\frac{V_{out}}{R_f} = -\frac{V_1}{R_1} - \frac{V_2}{R_2} - \ldots - \frac{V_n}{R_n}$ or $V_{out} = -\left(V_1\frac{R_f}{R_1} + V_2\frac{R_f}{R_2} + \ldots + V_n\frac{R_f}{R_n}\right)$.

3.2.5 *Simultaneous Addition and Subtraction*

A circuit simultaneously implementing addition and subtraction is depicted in Fig. 3.8. It contains $2n$ inputs with the corresponding voltages V_{1+}, V_{2+}, …, V_{n+} and V_{1-}, V_{2-},…,V_{n-} with $n \geq 1$. The currents passing to op-amp's inverting and non-inverting inputs are:

$$
\begin{aligned}
&\frac{V_{1+}-V_+}{R_{\text{in}}} = I_{1+}, && \frac{V_{1-}-V_-}{R_{\text{in}}} = I_{1-},\\
&\frac{V_{2+}-V_+}{R_{\text{in}}} = I_{2+}, && \frac{V_{2-}-V_-}{R_{\text{in}}} = I_{2-},\\
&\quad\vdots && \quad\vdots\\
&\frac{V_{n+}-V_+}{R_{\text{in}}} = I_{n+}, && \frac{V_{n-}-V_-}{R_{\text{in}}} = I_{n-},\\
&-\frac{V_+}{R_f} = I_{f+}, && \frac{V_{\text{out}}-V_-}{R_f} = I_{f-}.
\end{aligned} \tag{3.1}
$$

Kirchhoff's first law for the nodes A and B in Fig. 3.8 takes the forms $\sum_{i=1}^{n} I_{i-} + I_{f-} = 0$ and $\sum_{j=1}^{n} I_{j+} + I_{f+} = 0$. Substituting (3.1) into Kirchhoff's first law expressions, one obtains:

$$
V_+ = -\frac{R_f}{R_{\text{in}}} n V_+ + \frac{R_f}{R_{\text{in}}} \sum_{j=1}^{n} V_{j+}, \quad V_{\text{out}} = \frac{R_f}{R_{\text{in}}} n V_- + V_- - \frac{R_f}{R_{\text{in}}} \sum_{i=1}^{n} V_{i-}. \tag{3.2}
$$

Taking into account the condition $V_- = V_+$ (because the op-amp has negative feedback) and substituting the first expression for V_+ being equal to V_- into the second equation, one derives the output voltage formula:

$$
V_{\text{out}} = \frac{R_f}{R_{\text{in}}} \left(\sum_{j=1}^{n} V_{j+} - \sum_{i=1}^{n} V_{i-} \right). \tag{3.3}
$$

It is important to note that the circuit must have the same number of inputs connected to op-amp's inverting and non-inverting inputs. If you need to process $2n+1$ signals, then you must to create $2n + 2$ inputs ($n + 1$ to the inverting input and $n + 1$ to the non-inverting one) and to connect an unused input to ground.

3.2.6 *Functional Conversion of an Analog Signal (General Approach)*

Let us consider the circuit depicted in Fig. 3.9a where the op-amp feedback contains an element F. The functional dependence of the current I_F passing through the element F on the applied voltage V_F is described by the operator $\hat{f}$ such that $I_F = \hat{f}[V_F]$. In the particular case in Fig. 3.9a the op-amp input voltages are $V_- = V_+ = 0$ (because the op-amp has negative feedback). Then the currents through the resistor R and the element F are $I_R = V_{\text{in}}/R$ and $I_F = \hat{f}[-V_{\text{out}}]$. Kirchhoff's first law for the node A in Fig. 3.9a takes the form: $I_R = I_F$ (the op-amp input current I_- is

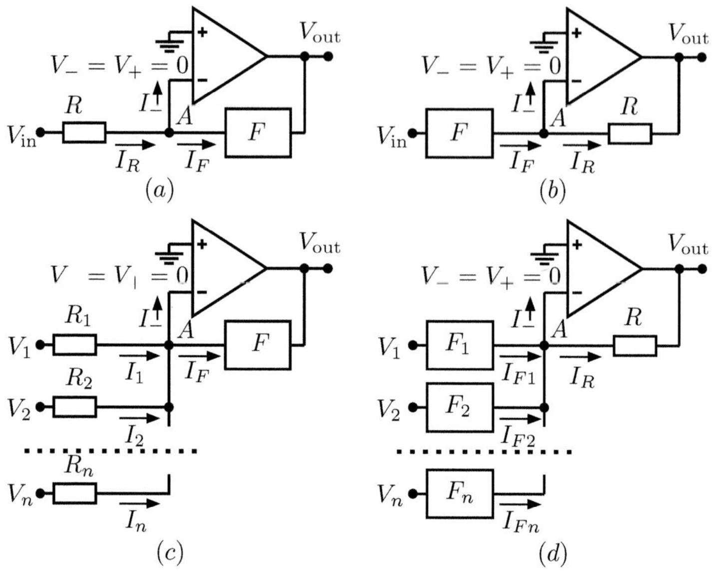

Fig. 3.9 Different options for functional conversion of a single signals (panels (**a**) and (**b**)) and a set of signals (panels (**c**) and (**d**))

neglected). Then $\frac{V_{\text{in}}}{R} = \hat{f}[-V_{\text{out}}]$ or in the inverse form:

$$V_{\text{out}} = -\hat{f}^{-1}\left[\frac{V_{\text{in}}}{R}\right], \tag{3.4}$$

where $\hat{f}^{-1}$ is an operator being inverse to the operator $\hat{f}$.

The circuit depicted in Fig. 3.9b represents the opposite configuration. Indeed, Kirchhoff's first law for the node A in Fig. 3.9b taking the form $I_F = I_R$ leads to the equality $\hat{f}[V_{\text{in}}] = -\frac{V_{\text{out}}}{R}$ or in the final form:

$$V_{\text{out}} = -R\hat{f}[V_{\text{in}}]. \tag{3.5}$$

The circuit in Fig. 3.9c has several inputs. In such a case Kirchhoff's first law for the node A is $I_1 + I_2 + \ldots + I_n = I_F$ or $\frac{V_1}{R_1} + \frac{V_2}{R_2} + \ldots + \frac{V_n}{R_n} = \hat{f}[-V_{\text{out}}]$. The inverse form is

$$V_{\text{out}} = -\hat{f}^{-1}\left[\frac{V_1}{R_1} + \frac{V_2}{R_2} + \ldots + \frac{V_n}{R_n}\right]. \tag{3.6}$$

The circuit being inverse to the functional converter in Fig. 3.9c is illustrated in Figs. 3.9d. It contains identical elements F_1, F_2, ..., F_n characterized by a functional current-voltage dependence $I_F = \hat{f}[V_F]$ and connected to the circuit

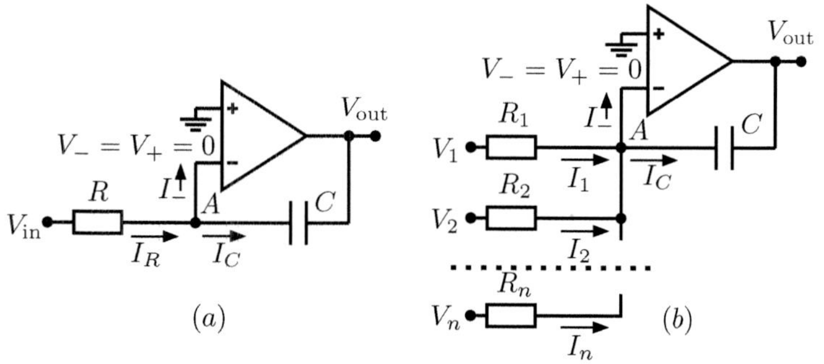

Fig. 3.10 Single-input (panel (**a**)) and multiple-input (panel (**b**)) op-amp-based analog integrators

inputs. Kirchhoff's first law for the node A is $I_{F1} + I_{F2} + \ldots + I_{Fn} = I_R$ or in the final form:

$$V_{\text{out}} = -R\left(\hat{f}[V_1] + \hat{f}[V_2] + \ldots + \hat{f}[V_n]\right). \tag{3.7}$$

Applications of the generalized circuits in Fig. 3.9 are considered in the further sections on particular examples.

3.2.7 *Inverting Integrator*

A single-input inverting integrator is shown in Fig. 3.10a. The circuit is a particular case of the functional converter in Fig. 3.9a with the capacitor C playing a functional element role. The capacitor current-voltage functional dependence is $I_C = \hat{f}[V_C] = C\frac{dV_C}{dt}$ which means that the functional operator is $\hat{f} = C\frac{d\ldots}{dt}$ and the inverse one becomes $\hat{f}^{-1} = \frac{1}{C}\int\limits_0^t \ldots dt$. Then the corresponding form of Eq. (3.4) becomes:

$$V_{\text{out}} = -\frac{1}{RC}\int\limits_0^t V_{\text{in}} dt. \tag{3.8}$$

In a case of a multiple-input inverting integrator (Fig. 3.10b) one deals with a particular circuit of the functional converter in Fig. 3.9c. Thus, the circuit equation takes the following form:

$$V_{\text{out}} = -\frac{1}{C}\int_0^t \left(\frac{V_1}{R_1} + \frac{V_2}{R_2} + \ldots + \frac{V_n}{R_n}\right) dt. \tag{3.9}$$

The discussed op-amp-based analog integrator is a central element of dynamical system analog models (see Sect. 3.3 where the analog modelling approach is discussed).

3.2.8 Inverting Differentiator

A single-input inverting differentiator is shown in Fig. 3.11a. The circuit is a particular case of the functional converter in Fig. 3.9b with the capacitor C playing a functional element role. The capacitor current-voltage functional dependence is $I_C = \hat{f}\,[V_C] = C\frac{dV_C}{dt}$. Then the corresponding form of Eq. (3.5) becomes:

$$V_{\text{out}} = -RC\frac{dV_{\text{in}}}{dt}. \tag{3.10}$$

In a case of a multiple-input inverting differentiator (Fig. 3.11b) one deals with a particular circuit of the functional converter in Fig. 3.9d. Thus, the circuit equation takes the following form:

$$V_{\text{out}} = -R\left(C_1\frac{dV_1}{dt} + C_2\frac{dV_2}{dt} + \ldots + C_n\frac{dV_n}{dt}\right). \tag{3.11}$$

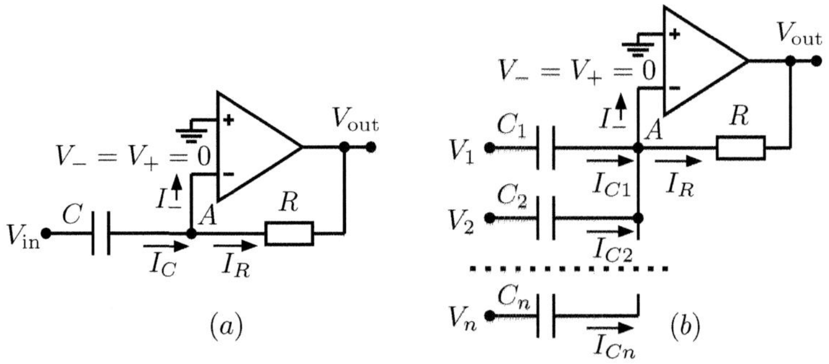

Fig. 3.11 Single-input (panel (**a**)) and multiple-input (panel (**b**)) op-amp-based analog differentiators

3.2.9 Logarithmic Amplifier

A single-input logarithmic amplifier is shown in Fig. 3.12a. The circuit is a particular case of the functional converter in Fig. 3.9a with the diode D playing a functional element role. The diode current-voltage functional dependence (schematically depicted in Fig. 2.10b) can be expressed as being $I_D = \hat{f}[V_D] = I_0\left(\exp\left(\frac{V_D}{n_{\text{id}}V_T}\right) - 1\right)$, where I_0 is the reverse bias saturation current, V_D is the voltage across the diode, $V_T = kT/q$ (k is the Boltzmann constant, T is an absolute temperature, q is the electron charge) is the thermal voltage, n_{id} is the ideality factor (also known as the quality factor or the emission coefficient). The diode voltage V_D is assumed to be positive. Since the thermal voltage V_T is a small quantity, the diode current-voltage characteristic can be simplified: $I_D = \hat{f}[V_D] = I_0 \exp\left(\frac{V_D}{n_{\text{id}}V_T}\right)$. Then the corresponding form of Eq. (3.4) becomes:

$$V_{\text{out}} = -n_{\text{id}} V_T \ln\left(\frac{V_{\text{in}}}{I_0 R}\right). \tag{3.12}$$

In a case of a multiple-input logarithmic amplifier (Fig. 3.12b) one deals with a particular circuit of the functional converter in Fig. 3.9c. Thus, the circuit equation takes the following form:

$$V_{\text{out}} = -n_{\text{id}} V_T \ln\left(\frac{1}{I_0}\left(\frac{V_1}{R_1} + \frac{V_2}{R_2} + \ldots + \frac{V_n}{R_n}\right)\right). \tag{3.13}$$

Transition to a new logarithm base can be achieved using the analog implementation of the formula $\log_b x = \log_a x / \log_a b$ corresponding to the output signal amplification or attenuation.

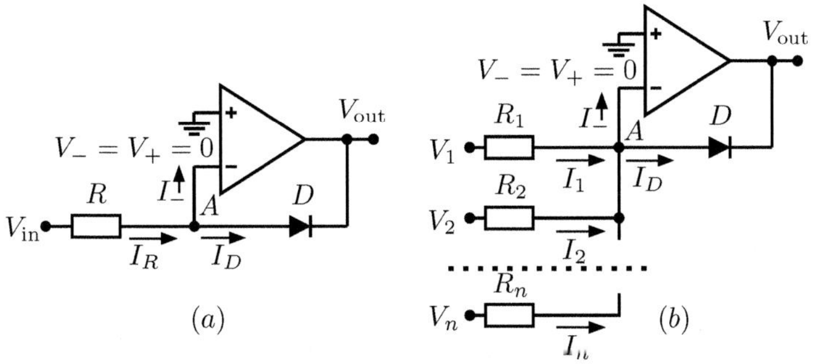

Fig. 3.12 Single-input (panel (**a**)) and multiple-input (panel (**b**)) op-amp-based logarithmic amplifiers

3.2.10 Exponential Amplifier

A single-input exponential amplifier is shown in Fig. 3.13a. The circuit is a particular case of the functional converter in Fig. 3.9b with the diode D playing a functional element role. The diode current-voltage functional dependence is assumed to be $I_D = \hat{f}\left[V_D\right] = I_0 \exp\left(\frac{V_D}{n_{\text{id}} V_T}\right)$. Then the corresponding form of Eq. (3.5) becomes:

$$V_{\text{out}} = -RI_0 \exp\left(\frac{V_{\text{in}}}{n_{\text{id}} V_T}\right). \tag{3.14}$$

In a case of a multiple-input exponential amplifier (Fig. 3.13b) one deals with a particular circuit of the functional converter in Fig. 3.9 d which contains a set of diodes assumed to be identical. Thus, the circuit equation takes the following form:

$$V_{\text{out}} = -RI_0 \left(\exp\left(\frac{V_1}{n_{\text{id}} V_T}\right) + \exp\left(\frac{V_2}{n_{\text{id}} V_T}\right) + \ldots + \exp\left(\frac{V_n}{n_{\text{id}} V_T}\right)\right). \tag{3.15}$$

If the diodes D_1, D_2,…,D_n have different characteristics, then the circuit output voltage formula becomes more complicated:

$$\begin{aligned} V_{\text{out}} = -R\Big(I_{01} \exp\left(\tfrac{V_1}{n_{\text{id1}} V_T}\right) + I_{02} \exp\left(\tfrac{V_2}{n_{\text{id2}} V_T}\right) \\ + \ldots + I_{0n} \exp\left(\tfrac{V_n}{n_{\text{idn}} V_T}\right)\Big). \end{aligned} \tag{3.16}$$

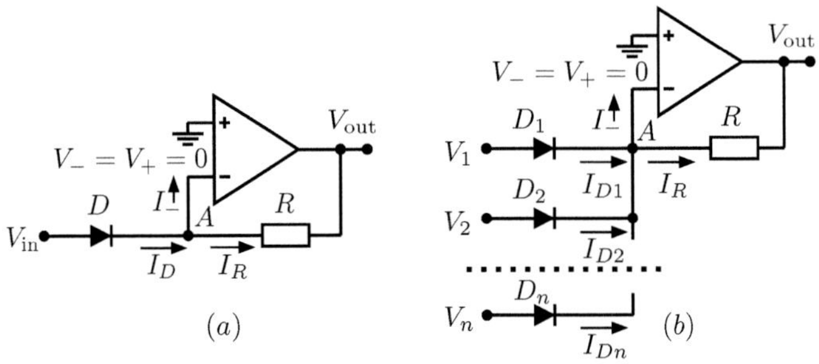

Fig. 3.13 Single-input (panel (**a**)) and multiple-input (panel (**b**)) op-amp-based exponential amplifiers

3.2.11 Multiplication

One of the possible ways for the multiplication circuit implementation involving special integrated circuits is discussed in Sect. 3.1.3. If multiplication of many signals is necessary or usage of integrated circuits is problematic, then one can use the circuit depicted Fig. 3.14 in a case of positive input voltages. The circuit contains a set of n logarithmic amplifiers based on the op-amps $U_{1,2,\ldots,n}$ the output voltages of which are $V_{\ln i} = -n_{\text{id}} V_T \ln\left(\frac{V_i}{I_0 R_{\text{in}}}\right)$ (see Sect. 3.2.9 for more details). Then the voltages $V_{\ln i}$ are summed by the amplifier U_s the output voltage of which becomes $V_s = -\sum_{i=1}^{n} V_{\ln i} = n_{\text{id}} V_T \sum_{i=1}^{n} \ln\left(\frac{V_i}{I_0 R_{\text{in}}}\right)$. The voltage V_s is the input signal of the exponential amplifier based on the op-amp U_{exp}. All the diodes in Fig. 3.14 are assumed to be identical. In such a case the circuit output voltage becomes:

$$V_{\text{out}} = -R_f I_0 \exp\left(\frac{V_s}{n_{\text{id}} V_T}\right) = -R_f I_0 \exp\left(\sum_{i=1}^{n} \ln\left(\frac{V_i}{I_0 R_{\text{in}}}\right)\right). \tag{3.17}$$

Using the formula $\ln(x_1) + \ln(x_2) + \ldots + \ln(x_n) = \ln(x_1 x_2 \ldots x_n)$, one obtains the final expression for the circuit output voltage:

$$V_{\text{out}} = -R_f I_0 \exp\left(\ln\left(\frac{V_1 V_2 \ldots V_n}{(I_0 R_{\text{in}})^n}\right)\right) = -\frac{R_f I_0}{(I_0 R_{\text{in}})^n} V_1 V_2 \ldots V_n. \tag{3.18}$$

In the simplest case $R_f = R_{\text{in}} = 1/I_0$ the output voltage becomes $V_{\text{out}} = V_1 V_2 \ldots V_n$. The circuit in Fig. 3.14 works correctly when expected voltage values at any points of the circuit do not exceed the power supply range $[V_-, V_+]$.

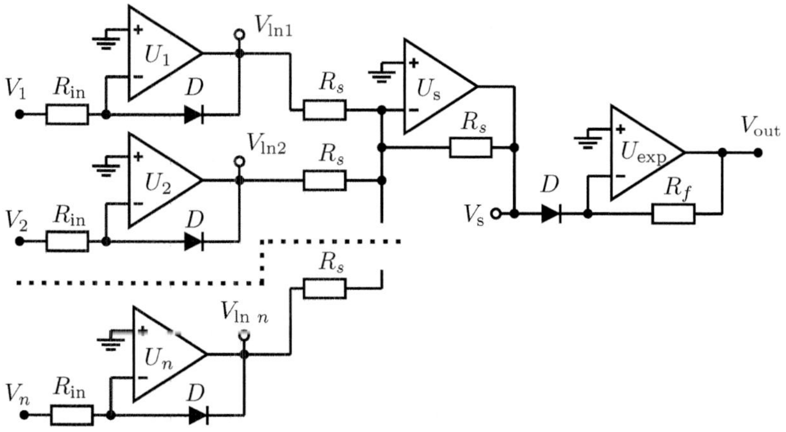

Fig. 3.14 Multiple-input multiplying circuit. Input voltages are assumed to be positive, $V_{1,2,\ldots,n} > 0$

3.2.12 Division

Similarly to the multiplication implementation discussed in the previous section, logarithmic-exponential signal transformations allow to realize a single-quadrant division procedure (Fig. 3.15a). In contrast to Fig. 3.14, the circuit in Fig. 3.15a contains the differential amplifier U_s the output voltage of which is $V_s = V_{\ln 1} - V_{\ln 2}$. All the diodes in Fig. 3.15a are assumed to be identical. Then the voltage V_s becomes: $V_s = n_{id} V_T \left(\ln\left(\frac{V_2}{I_0 R_{in}}\right) - \ln\left(\frac{V_1}{I_0 R_{in}}\right)\right)$. Finally, one obtains the circuit output voltage $V_{out} = -R_f I_0 \exp(V_s) = -R_f I_0 \exp\left(\ln\left(\frac{V_2}{I_0 R_{in}}\right) - \ln\left(\frac{V_1}{I_0 R_{in}}\right)\right) = -R_f I_0 \exp\left(\ln\left(\frac{V_2}{V_1}\right)\right) = -R_f I_0 \frac{V_2}{V_1}$.

The circuit in Fig. 3.15a works only when the input voltages $V_{1,2}$ are positive. If the voltages $V_{1,2}$ can be positive and negative, then one must apply the second approach illustrated in Fig. 3.15b, c. Here the op-amp has a feedback loop containing a multiplication block (single multiplier or a multiplier-based circuit). These circuits are similar to inverting and non-inverting amplifiers discussed in Sect. 3.2.2 except of the presence of the multiplier in the feedback loop. The multiplier output voltage is assumed to be $V_f = \frac{V_{out} V_2}{V_{ref}}$, where V_{ref} is a constant used for the output voltage scaling. Since an op-amp tends to eliminate the voltage difference between the inverting and non-inverting inputs, one can describe the circuit in Fig. 3.15b by the equality $V_f = V_1 = \frac{V_{out} V_2}{V_{ref}}$ or $V_{out} = V_{ref} \frac{V_1}{V_2}$.

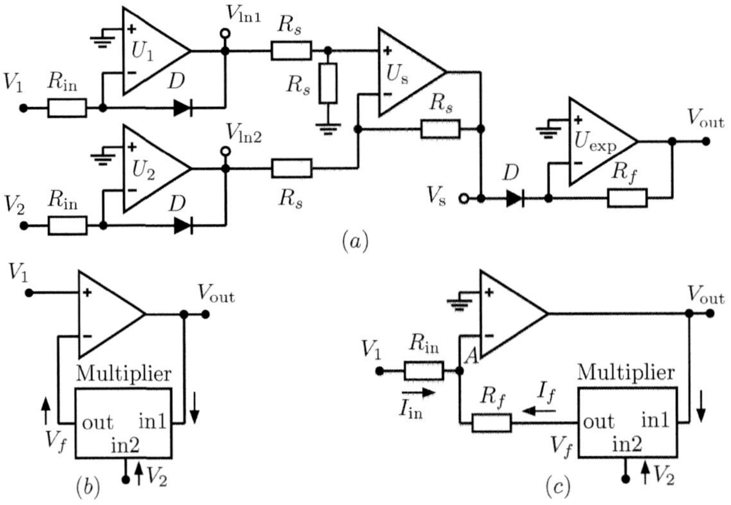

Fig. 3.15 (**a**) Op-amp based single-quadrant divider. Input voltages are assumed to be positive, $V_{1,2} > 0$; (**b**), (**c**) Non-inverting (panel (**b**)) and inverting (panel (**c**)) four-quadrant divider which contains a multiplier in an op-amp feedback loop. The multiplier output voltage is $V_f = \frac{V_{out} V_2}{V_{ref}}$

Similarly, the condition $V_- = V_+ = 0$ takes place in a case of the inverting divider in Fig. 3.15c. Then the current I_{in} flowing through the resistor R_{in} becomes: $I_{in} = V_1/R_{in}$. Similarly, the current I_f passing through the resistor R_f equals to V_f/R_f. Kirchhoff's first law for the node A in Fig. 3.15c takes the form $I_{in} = -I_f$ (because the op-amp input current is supposed to be zero). Substituting formulas for currents, one obtains: $\frac{V_1}{R_{in}} = -\frac{V_{out}V_2}{R_f V_{ref}}$ or $V_{out} = -V_{ref}\frac{R_f}{R_{in}}\frac{V_1}{V_2}$.

In addition, there exist analog integrated circuits including both an op-amp and a multiplier (for instance, different modifications of the AD534 integrated circuit) which allow to simply implement division of two analog signals with high accuracy. Depending on the connection specifics, one can realize different operations (multiplication, division, etc.) by using such integrated circuits (see the corresponding datasheets).

3.2.13 Simultaneous Multiplication and Division

The circuit solutions considered in the previous two sections can be improved for instantaneous realization of two procedures: multiplication and division of three or more analog signals. In a case of the circuits including logarithmic and exponential signal transformations (Figs. 3.14 and 3.15a) it implies change of a summing or differential amplifier U_s by the generalized circuit considered in Sect. 3.2.5. The circuit in Fig. 3.16a contains the amplifier U_s implementing simultaneous addition and subtraction the output voltage of which is $V_s = n_{id}V_T\left(\ln\left(\frac{V_1}{I_0 R_{in}}\right) - \ln\left(\frac{V_2}{I_0 R_{in}}\right) - \ln\left(\frac{V_3}{I_0 R_{in}}\right)\right)$. Finally, one obtains the circuit output voltage $V_{out} = -R_f I_0 \exp(V_s) = -R_f I_0 \exp\left(\ln\left(I_0 R_{in}\frac{V_1}{V_2 V_3}\right)\right) = -R_f R_{in} I_0^2 \frac{V_1}{V_2 V_3}$.

Circuits implementing simultaneous four-quadrant multiplication and division are similar to ones depicted in Fig. 3.15b, c except of multiplying block in the op-amp feedback. The multipliers in Fig. 3.16b, c create the output voltage $V_f = \frac{V_{out}V_2 V_3}{V_{ref}}$. Then the output voltage of the circuit in Fig. 3.16b is $V_{out} = V_{ref}\frac{V_1}{V_2 V_3}$ while the output voltage of the circuit in Fig. 3.16c is $V_{out} = -V_{ref}\frac{R_f}{R_{in}}\frac{V_1}{V_2 V_3}$.

3.2.14 Functions $f(x) = x^a$, $f(x) = a^x$ and $f(x, y) = x^y$

In a case of an integer power factor the function $f(x) = x^a$ can be simply implemented by using a single multiplier or divider as well as by a cascade of ones. In a case of the real non-integer factor a one can obtain the necessary function by means of logarithmic-exponential transformations as in Fig. 3.17a. Here the signal V_{in} is initially transformed into $V_{ln} = -n_{id}V_T \ln\left(\frac{V_{in}}{I_0 R_{in}}\right)$. After that the

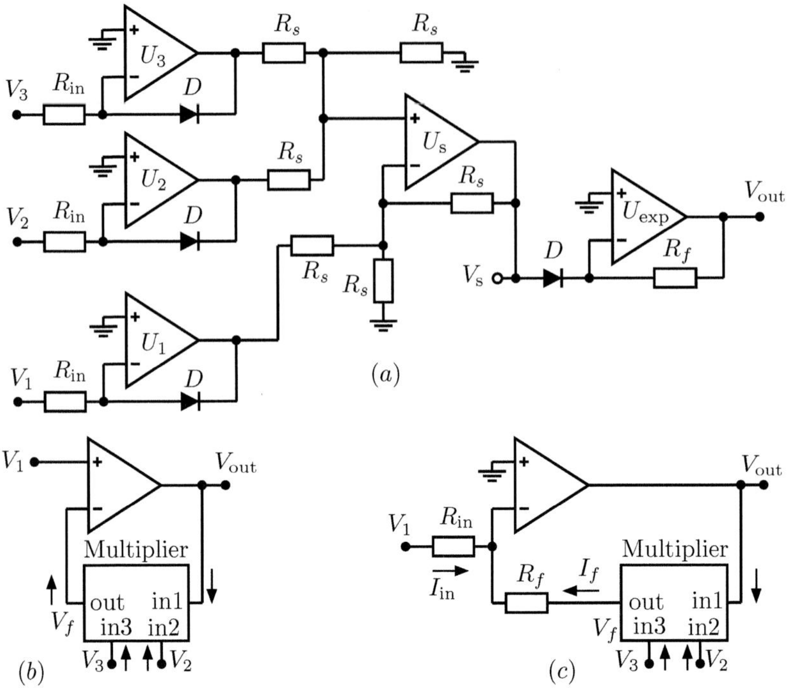

Fig. 3.16 (**a**) Op-amp based single-quadrant circuit for simultaneous multiplication and division. Input voltages are assumed to be positive, $V_{1,2,3} > 0$; (**b**), (**c**) Non-inverting (panel (**b**)) and inverting (panel (**c**)) circuits implementing four-quadrant multiplication and division. The output voltage of the op-amp feedback multiplier is $V_f = \frac{V_{out}V_2V_3}{V_{ref}}$

inverting amplifier produces the signal $V_m = -\frac{R_2}{R_1}V_{ln} = \frac{R_2 n_{id} V_T}{R_1} \ln\left(\frac{V_{in}}{I_0 R_{in}}\right) = n_{id} V_T \ln\left(\left(\frac{V_{in}}{I_0 R_{in}}\right)^a\right)$ where $a = R_2/R_1$. The exponential amplifier based on the op-amp U_{exp} produces the circuit output voltage $V_{out} = -I_0 R_f \exp\left(\ln\left(\left(\frac{V_{in}}{I_0 R_{in}}\right)^a\right)\right) = -I_0 R_f \left(\frac{V_{in}}{I_0 R_{in}}\right)^a$.

Change of the inverting amplifier by a multiplier (Fig. 3.17b) allows to realize the function $f(x, y) = x^y$ if the multiplier output voltage is $V_m = -V_{ln}V_2/V_{ref}$ where V_{ref} is a scaling fixed constant. Then the multiplier output voltage becomes $V_m = \frac{V_2}{V_{ref}} n_{id} V_T \ln\left(\frac{V_1}{I_0 R_{in}}\right)$. Finally, the circuit output voltage takes the form $V_{out} = -I_0 R_f \exp\left(\ln\left(\left(\frac{V_{in}}{I_0 R_{in}}\right)^a\right)\right) = -I_0 R_f \left(\frac{V_{in}}{I_0 R_{in}}\right)^a$, where $a = V_2/V_{ref}$. If the voltage V_1 is constant, then one deals with the function $f(x) = a^x$. It is important to re-note that all the diodes in Fig. 3.17a, b are assumed to be identical.

The input voltages V_{in} and V_1 in Fig. 3.17a, b are positive. However, the domain of the function $f(x) = \sqrt[a]{x}$ is $x \in (-\infty, +\infty)$ in a case of a positive odd number

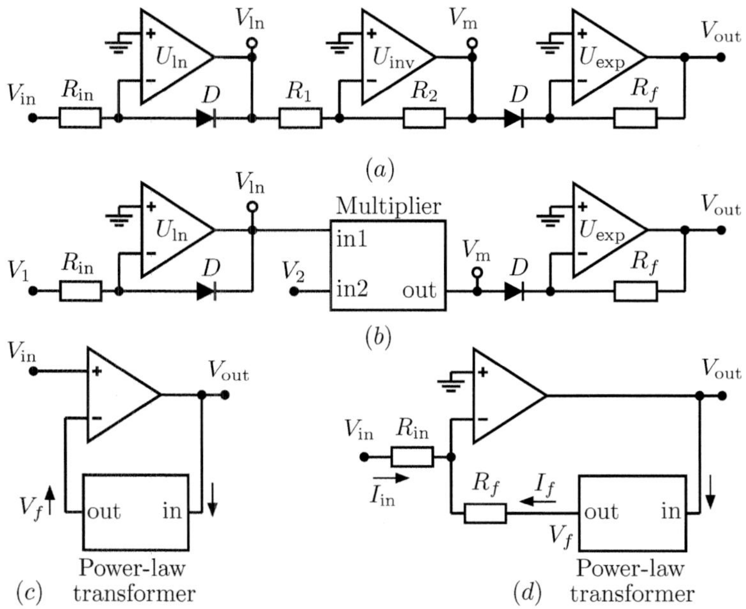

Fig. 3.17 Electronic implementation of the functions $f(x) = x^a$ (panel (**a**)) and $f(x, y) = x^y$ (panel (**b**)). The output voltage of the multiplier in panel (**b**) is $V_m = -\frac{V_{\ln} V_2}{V_{ref}}$; (**b**), (**c**) Non-inverting (panel (**b**)) and inverting (panel (**c**)) circuits implementing raising to a power. The output voltage of the op-amp feedback power-law transformer is $V_f = \frac{V^n_{out}}{V_{ref}}$, where a is an integer constant

a and $x \in (-\infty, 0) \cup (0, +\infty)$ in a case of a negative odd number a. To realize the function $f(x) = \sqrt[a]{x}$ in the full domain, one can use an operational amplifier with a feedback loop including four-quadrant multiplier- or divider-based power-law transformer (Fig. 3.17c, d). The output voltage of the power-law transformer is $V_f = \frac{V^a_{out}}{V_{ref}}$, where a is an integer constant. Then the output voltage of the non-inverting circuit (Fig. 3.17c) is $V_{out} = V_{ref}\sqrt[a]{V_{in}}$. The inverting circuit output voltage (Fig. 3.17d) becomes $V_{out} = -\frac{R_f}{R_{in}} V_{ref}\sqrt[a]{V_{in}}$.

3.2.15 Piecewise-Smooth Functions

Piecewise-smooth functions can be implemented by applying two methods. The first one implies an op-amp-based inverting half-wave rectifier. A particular circuit of the inverting half-wave rectifier is depicted in Fig. 3.18a. Generally, the circuit works as an inverting op-amp-based amplifier except of the negative feedback which consists of two branches. It is assumed that the diode conductance possess two values depending on the applied voltage sign: zero or infinity. As a result, either feedback branch carries an electric current while the second one blocks the current

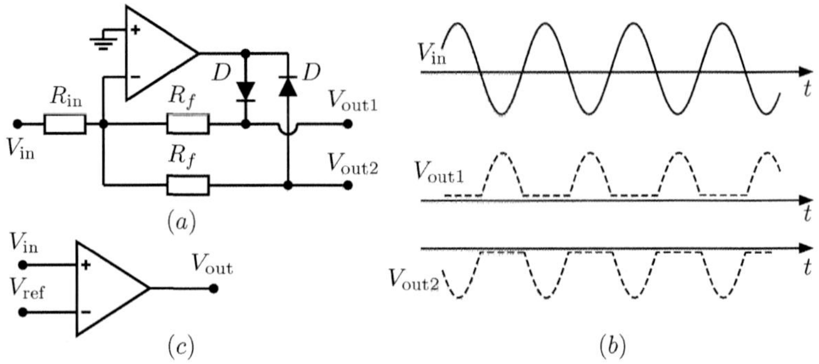

Fig. 3.18 Op-amp-based inverting half-wave rectifier (panel (**a**)) and particular input-output time realizations (panel (**b**)); (**c**) Analog comparator

flow. Since the circuit represents a modification of the inverting amplifier, its output voltages are:

$$V_{out1} = \begin{cases} -V_{in}, & V_{in} < 0, \\ 0, & V_{in} \geq 0, \end{cases} \qquad V_{out2} = \begin{cases} 0, & V_{in} < 0, \\ -V_{in}, & V_{in} \geq 0. \end{cases} \tag{3.19}$$

The circuit output signals in a case of an alternating input voltage are depicted in Fig. 3.18b. The circuit in Fig. 3.18a is practically used in Sect. 5.4.2.

The second way to implement piecewise-smooth functions is using an analog comparator. The output voltage of the comparator depicted in Fig. 3.18c is described by the formula:

$$V_{out} = \begin{cases} a, & V_{in} < V_{ref}, \\ b, & V_{in} \geq V_{ref}. \end{cases} \tag{3.20}$$

An operational amplifier without feedback can be used as a comparator. However, characteristics of a saturated operational amplifier can be unsatisfactory. In such a case one can use comparator integrated circuits. For example, the comparator integrated circuit AD790JNZ is used in Sect. 5.5.2 for the implementation of Chua's memristor with a piecewise-smooth characteristic.

3.2.16 Absolute Value

There are at least two options for extracting an input voltage absolute value. The first one is to use two blocks implementing the functions x^2 and $\sqrt{x}$ and connected in series as in Fig. 3.19a. The second technics implies the circuit in Fig. 3.18a extended as in Fig. 3.19b. Here the half-wave rectifier based on the op-amp U_r produces the signal V_r possessing two values: $V_r = -V_{in}$ for $V_{in} < 0$ and $V_r = 0$ for $V_{in} \geq 0$.

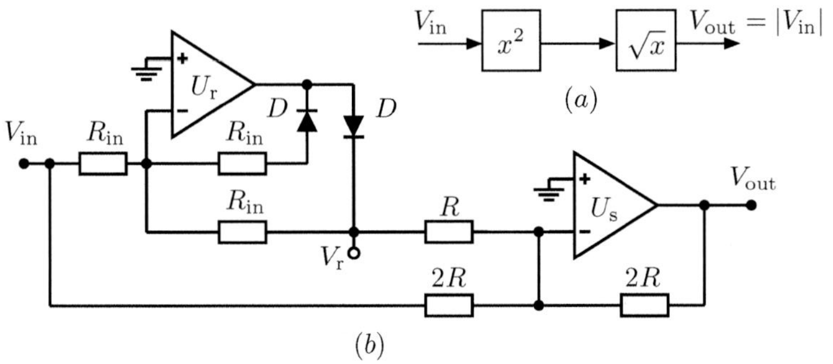

Fig. 3.19 Schematic illustration of the absolute value extraction including the operations of squaring and square rooting; (**b**) The circuit with an inverting half-wave rectifier extracting the absolute value of the input signal: $V_{out} = -|V_{in}|$

Then the signal V_r comes to the input of the summing amplifier U_s. Finally, the circuit output voltage becomes:

$$
\begin{aligned}
V_{out} &= -(2V_r + V_{in}) = \begin{cases} -(-2V_{in} + V_{in}), & V_{in} < 0, \\ -V_{in}, & V_{in} \geq 0, \end{cases} \\
&= \begin{cases} V_{in}, & V_{in} < 0, \\ -V_{in}, & V_{in} \geq 0, \end{cases} = -|V_{in}|.
\end{aligned}
\tag{3.21}
$$

3.2.17 *Trigonometric Functions*

The first approach for the trigonometric function implementation is based on the Taylor expansion. For instance, the Taylor series well-approximating the function $\sin(x)$ in the range $x \in [-180° : 180°]$ can be taken as the seventh-order polynomial: $\sin(x) \approx x - \frac{x^3}{3!} + \frac{x^5}{5!} - \frac{x^7}{7!}$. The analog circuit realization for such polynomial will consist of a large number of operational amplifiers, analog multipliers and other elements. In such a case, it seems easier and more rational to use universal trigonometric converters developed as integrated circuits. For instance, one can use converter AD639 by Analog Devices, which transforms the input voltage into its sine, cosine, tangent, cotangent, secant, cosecant as well as their inverse functions for a wide range of angles (up to $\pm$ 500°, see the datasheet) associated with the input voltage signals.

3.3 Analog Modelling of Dynamical Systems

The principles of analog modelling of deterministic and stochastic dynamical systems are well-described in review [4]. In Chap. 3, the only basic aspects are discussed. Many of interesting solutions and techniques developed by the author of this book for efficient and smart electronic realizations are demonstrated in Chap. 5 on concrete examples of dynamical systems.

3.3.1 One-Dimensional Systems

Here, we begin to discuss an alternative approach for dynamical system implementations by means of electronics. In contrast to the classical approach where dynamical variables are voltages and currents, analog modelling implies all the dynamical variables to be associated with integrated circuit output voltages. The description of such approach starts from the simplest task. Suppose one needs to implement the equation $\dot{x} = f(x)$. For this purpose, one can use the inverting integrator described in Sect. 3.2.7. It is assumed that the inverting integrator output voltage marked as signal x in Fig. 3.20a represents the instantaneous state of the dynamical variable x at any time t. Then signal x is processed by the functional block such that its output signal is a function $-f(x)$ of the input x. Finally, voltage $-f(x)$ returns to the integrator as the input signal and the integrator equation takes the form $x = -\frac{1}{RC}\int_0^t -f(x)dt$. Extracting the time derivative of the left and right parts of the equality, one obtains the modified formula for the integrator functioning: $RC\frac{dx}{dt} = f(x)$. Finally, a new time variable is introduced as $\tau = \frac{1}{RC}t$. Then the time derivative is recalculated with taking into account the new time scale: $\frac{dx}{d\tau} = \frac{dx}{dt}\frac{dt}{d\tau} = \frac{dx}{dt}RC$. Substituting this, the integrator's input-output relationship can be rewritten: $\frac{dx}{d\tau} = f(x)$ or in the simplified form $\dot{x} = f(x)$.

Using the same approach, more complicated one-dimensional ordinary differential equations can be implemented. For instance, the circuit depicted

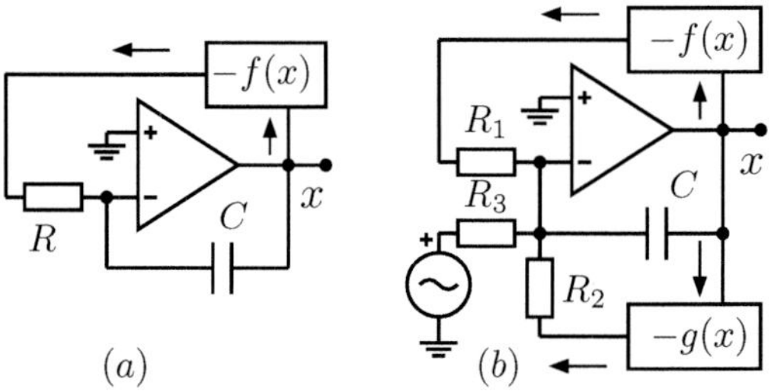

Fig. 3.20 Electronic implementation of one-dimensional ordinary differential equations by means of analog modelling

in Fig. 3.20b contains two independent functional blocks and the external periodic forcing. In such a case, the integrator's equation becomes $x = -\frac{1}{C}\int_0^t\left(-\frac{f(x)}{R_1} - \frac{g(x)}{R_2} + \frac{A\sin(\omega t)}{R_3}\right)dt$. Extracting the time derivative, one rewrites the equation: $R_1 C\frac{dx}{dt} = f(x) + \frac{R_1}{R_2}g(x) - \frac{R_1}{R_3}A\sin(\omega t)$. After rescaling the time, $\tau = \frac{1}{R_1 C}t$, the equation takes the final form: $\frac{dx}{d\tau} = f(x) + \frac{R_1}{R_2}g(x) - \frac{R_1}{R_3}A\sin(R_1 C\omega\tau)$. It is important to note that tuning the resistances $R_{1,2,3}$ allows to adjust the gain coefficients at functions $g(x)$ and $A\sin(R_1 C\omega\tau)$ (here, the author means the factors $\frac{R_1}{R_2}$ and $\frac{R_1}{R_3}$) without additional attenuators or integrated-circuit-based amplifiers. It provides for reduction of an electronic component number and for simplification of dynamical system analog model.

3.3.2 *Multidimensional Systems*

Consider the circuit depicted in Fig. 3.21. The circuit contains two integrators which output voltages x and y are assumed to be dynamical variables of the developed analog model. These signals are used as inputs of the blocks $-f(x, y)$ and $-g(x, y)$ which produce the input signals for the integrators. The equations describing the integrators are $R_x C_x\frac{dx}{dt} = f(x, y)$ and $R_y C_y\frac{dy}{dt} = g(x, y)$. After the time rescaling, $\tau = \frac{1}{R_x C_x}t$, one obtains the system of dynamical equations in the final form:

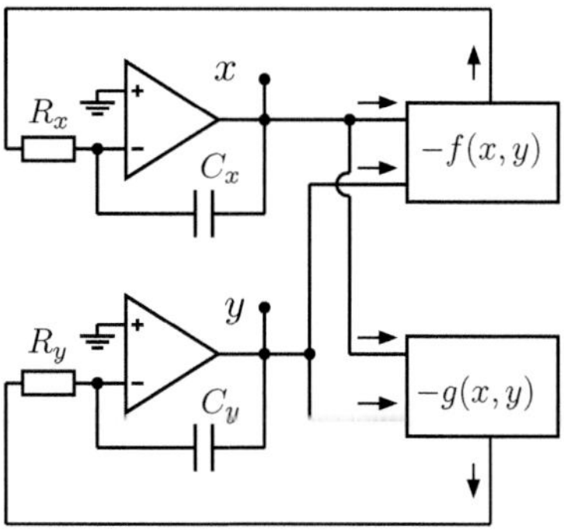

Fig. 3.21 Electronic implementation of a two-dimensional dynamical system by means of analog modelling

$$\begin{cases} \dfrac{dx}{d\tau} = f(x, y), \\ \\ \dfrac{R_y C_y}{R_x C_x} \dfrac{dy}{d\tau} = g(x, y). \end{cases} \tag{3.22}$$

Varying the values $R_{x,y}$ and $C_{x,y}$ one can implement a system of ordinary differential equations for dynamical variables evolving in different scales (see, for instance, the slow-fast dynamics). This approach is used for the implementation of the FitzHugh-Nagumo model discussed in Sect. 5.3.3.

In the similar way, one can use three or more integrators for the implementation of three-dimensional or multidimensional dynamical systems. Here, the number of integrators determines the dimensionality of the dynamical system realized in electronic experiments.

3.3.3 *Practical Recommendations and Frequent Technical Problems*

Advantages and disadvantages of certain analog models are considered in Chap. 5. In this section, the practical issues of analog modelling are summarised and discussed in a general form, but on examples of concrete circuits. Please analyse all the bullet points below before you start to develop your analog model of any dynamical system. In addition, it must be useful to study practical recommendations given in Ref. [4].

Analyse All the Expected Oscillation Amplitudes Before Starting the Circuit Development

The main principle you must keep in your mind: the output voltage of any integrated circuit is finite and cannot exceed the voltage supply. All the operational amplifiers and analog multipliers discussed in this book are connected to the power supply $\pm$ 15 V. Then you must be absolutely sure that the voltage at any point of the circuit is smaller that the voltage supply.

Suppose that the dynamical model you need to realize in electronic experiments exhibits the oscillations characterised by the amplitude $x_{\max} = 25$ V. The power supply voltage is $\pm$ 15 V. If you directly implement the system under study, the maximal achieved voltage amplitude is $\pm$ 15 V and the developed circuit will not operate correctly. In such a case, it is useful to rewrite the system equations for the dynamical variable $y = x/2$. Then the expected oscillation amplitude becomes $y_{\max} = 12.5$ V which does not exceed the supply voltage. Usually, developers prefer to simulate the dynamical systems in numerical experiments (by means of

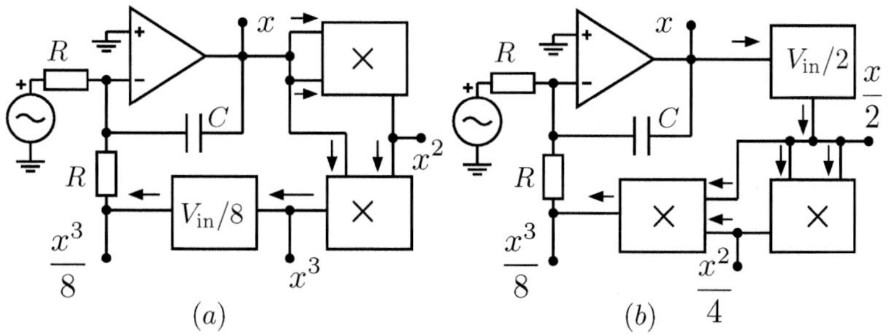

Fig. 3.22 Analog models for the dynamical system $\dot{x} = -\frac{1}{8}x^3 - A\sin(\omega t)$ for $x(t) \in [-3 : 3]$ V: the wrong circuit (panel (**a**)) and the correct one (panel (**b**))

integration methods) before the electronic circuit implementations to estimate the expected amplitudes.

Voltage Signal at Any Point of a Circuit Cannot Exceed the Power Supply Voltage

Suppose you need to implement the dynamical system $\dot{x} = -\frac{1}{8}x^3 - A\sin(\omega t)$ (this model is just an example and does not refer to any known dynamical system). It is expected that the external forcing amplitude A is chosen such that the dynamical variable oscillates in the range $x \in [-3 : 3]$ V. Figure 3.22 schematically illustrates two possible solutions to implement the dynamical system. The circuit in Fig. 3.22a will not work correctly when instantaneous values $x(t)$ tend to $\pm$ 3 V. This is due to the fact that for $x(t) = \pm 3$ V the necessary voltage at terminal A is $x^3 = \pm 27$ V. However, the real voltage maximum at this point is $\pm$ 15 V (it is bounded by power supply voltage). This problem is avoided in Fig. 3.22b where the signal $0.5x$ is extracted before the cubic nonlinearity implementation. In such a case, one deals with signals $0.25x^2 \in [-2.25 : 2.25]$ V and $x^3/8 \in [-3.375 : 3.375]$ V which do not exceed the voltage of power supply.

Avoid Extremely Low Amplitudes as Well as Extremely High Amplification or Attenuation

If a voltage signal amplitude at any terminal of your circuit is extremely low and this signal is amplified such that the gain is sufficiently high, then the action of noises and inaccuracies can be more significant. In addition, the impact of the mentioned negative factors can be enhanced if any circuit block realizes division by an extremely weak signal. It is recommended to avoid such situations by reconsideration of the model under study in new variables (see the previous bullet

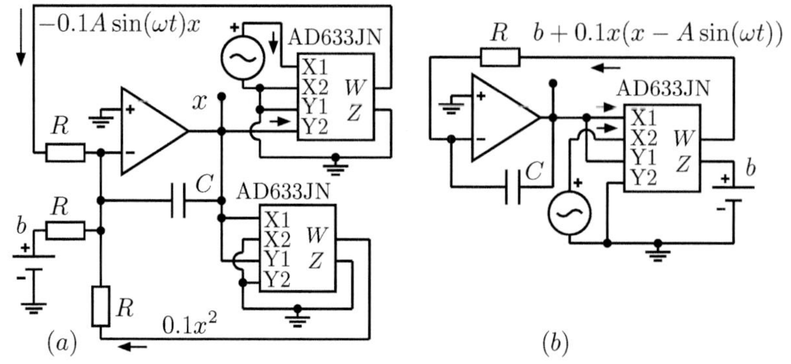

Fig. 3.23 Analog models for the dynamical system $\dot{x} = -b + 0.1(A\sin(\omega t) - x)x$. The circuit in panel (**a**) is too complicated, while the circuit in panel (**b**) is simplified

points). This approach is described in details in Sect. 5.6 on an example of a concrete delayed-feedback oscillator.

Minimize a Number of Circuit Elements as Much as Possible

Figure 3.23 illustrates two circuits implementing the same dynamical model $\dot{x} = -b + 0.1(A\sin(\omega t) - x)x$. The circuit in Fig. 3.23a realizes the system equation in a more complicated form as $\dot{x} = -b + 0.1A\sin(\omega t)x - 0.1x^2$ and includes three resistors and two analog multipliers AD633JN. In contrast, the same equation can be implemented by using less number of elements as depicted in Fig. 3.23b. The circuit in Fig. 3.23b realises the considered model in the form $\dot{x} = -b - 0.1(x - A\sin(\omega t))x$. The less is the number of circuit elements the higher is the circuit reliability. In addition to reduced financial costs and power consumption, the circuit simplification provides for reducing the impact of noises and inaccuracies.

The Best Resistor Values Are Within the Range 1–100 kΩ, While the Best Capacitance Values Must be in the Range 1–100 nF

Similar values for analog model circuit elements are recommended in review [4] and are dictated by practical experience. The lower are resistor and capacitor values at the inverting integrator, the higher is the main oscillation frequency of the analog model circuit. If it is impossible to avoid high frequencies, then the used operational amplifier must be chosen with taking into account an appropriate input signal frequency and operational amplifier's slew rate.

3.4 Operational-Amplifier-Based Implementation of Two-Terminal Elements

Operational amplifiers are widely used for the implementations of two-terminal elements with various characteristics. The operational amplifier is the central element of the circuits briefly discussed in this section. In addition to these mentioned examples, operational-amplifier-based blocks can be included into more complicated circuits emulating two-terminals to realize nonlinear signal transformations. Such approach is demonstrated in Chap. 5 on an example of Chua's memristor model with piecewise-smooth characteristics based on a transistor switch (see Sect. 5.5.1).

3.4.1 *Gyrator*

Usually, a gyrator is used to transform a load capacitance into an inductance. If one develops a circuit working at low frequencies and deals with low load powers, the behaviour of the gyrator can be reproduced by the operational-amplifier-based circuit depicted in Fig. 3.24a. This allows to implement an inductive element in a small electronic circuit or integrated circuit. The equivalent circuit for the discussed operational-amplifier-based circuit is illustrated in Fig. 3.24b: the circuit operation results in an input impedance of the form of an ideal inductor $L = R_L RC$ with a series resistance R_L, $Z_{\text{in}} = R_L + j\omega R_L RC$. The significance of such approach consists in the ability to create a two-terminal circuit being an analog of big coils of wire with large inductance. Thus, an inductor can be replaced by a much smaller assembly containing a capacitor, operational amplifiers or transistors, and resistors, which is especially useful in integrated circuit technologies. Nevertheless, it is important to note that the gyrators are active electronic circuits and cannot replace physical inductors in all the possible applications as they do not possess all the unique properties of physical inductors.

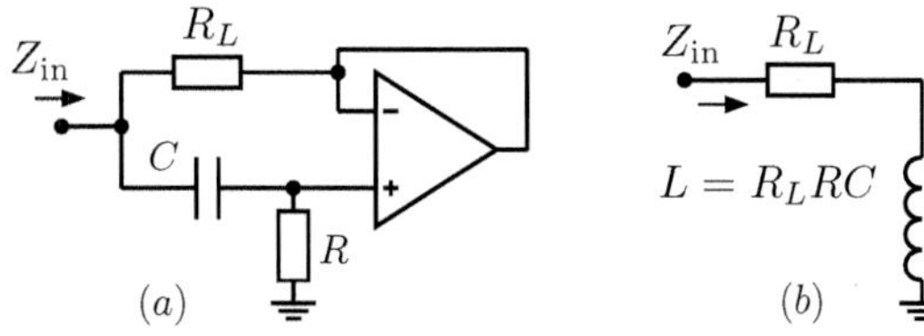

Fig. 3.24 Gyrator simulating inductance: circuit implementation based on an operational amplifier (panel (**a**)) and its equivalent circuit scheme (panel (**b**))

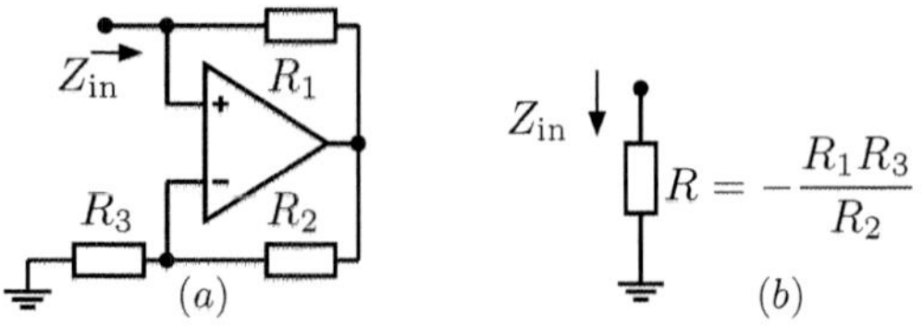

Fig. 3.25 Negative impedance converter: circuit implementation based on an operational amplifier (panel (**a**)) and its equivalent circuit scheme (panel (**b**))

3.4.2 Negative Impedance Converter

An example of the circuit implementing a negative impedance (also called negative impedance converter) is depicted in Fig. 3.25a. In contrast to an ordinary resistive load that consumes energy from a source circuit, the negative impedance circuit is an active circuit block and produces energy into circuits. The equivalent circuit scheme for the converter in Fig. 3.25a is a resistor characterised by a negative resistance value, $R = -\frac{R_1 R_3}{R_2}$ (see Fig. 3.25b). In general, elements $R_{1,2,3}$ do not need to be pure resistances (i.e., they may be capacitors, inductors, or impedance networks).

References

1. P. Horowitz, W. Hill, *The Art of Electronics*, 3rd edn. (Cambridge University Press, Cambridge, 2015)
2. W. Jung, *Op Amp Application Handbook* (Newnes, Oxford, 2004)
3. B. Carter, R. Mancini, *Op Amps for Everyone*, 5th edn. (Newnes, Oxford, 2017)
4. D. Luchinsky, P. McClintock, M. Dykman, Rep. Prog. Phys. **61**(8), 889 (1998)

Chapter 4
Special Issues of Electronic Modelling

In this chapter, we discuss specific issues which do not directly relate to electronic modelling methods discussed above, but could be useful in particular cases. A manifold of considered problems includes combination of analog and digital electronics in the same setups, as well as using a personal computer as an experimental setup element and noise generation.

4.1 Time Delay Implementation

Time delay is inherently present in a wide range of systems in physics, biology, economics, and other fields [1]. In particular, a manifold of biological processes associated with time delay ranges from individual [2] and collective [3] neural cell activity to the population dynamics [4, 5]. A large variety of applications associated with natural delay and imposed time-delay processes includes medical issues such as therapy for Parkinson's disease management [6–8] and control of other pathological brain rhythms (for instance, epileptic seizures) [9, 10], cardiac rhythm control [11], epidemic propagation control [12, 13]. In addition, time-delay processes play a significant role in chemical reaction rates [14, 15], laser dynamics [16–19], analog and digital electronic circuits [20, 21], traffic flow simulation [22], climate change modelling [23].

Besides the presence of delays in biological neural networks and their influence on the neural network behaviour [3, 24–27], I would like to emphasize the impact of time-delay loops in the context of artificial intelligence. Time delays are widely used for the development of delay-based machine learning algorithms involving spiking neural networks [28, 29]. Long time-delay loops provide for physical realization of reservoir computing architecture [30, 31]. The introduction of time-delay reservoir computing enabled simple optical, electronic and optoelectronic hardware implementations, which led to improvements of computation

© The Author(s), under exclusive license to Springer Nature Switzerland AG 2024

V. Semenov, *Electronic Modelling of Deterministic and Stochastic Oscillators*,
Springer Series in Synergetics, https://doi.org/10.1007/978-3-031-81553-9_4

time scales for supervised learning [32–35]. Moreover, deep-learning architectures have been adopted for the reservoir computing [36–38]. The delay-based reservoirs were successfully applied to a wide range of tasks, such as chaotic time series forecasting or speech recognition. In addition, using a single bistable delayed-feedback dynamical node as a network, one can synthesize a physical device (for instance, see the delayed-feedback-based Ising machine [39]) for solving problems with non-deterministic polynomial-time hardness (so-called NP-hard problems).

In the context of nonlinear dynamics, time delay represents a factor being responsible for a broad spectrum of fundamental phenomena such as delay-induced regular oscillatory behavior [40–44], delay-induced bifurcations [45, 46] and multistability [47, 48], stabilization of steady states and periodic orbits [49, 50], vibrational resonance [51, 52], delay-induced chaos [53, 54], delayed-feedback control of chaos [49] (including spatiotemporal patterns [55, 56]), delay-induced synchronization [57–59] and desynchronization [60, 61]. In addition to the deterministic effects, time delay represents a useful approach for controlling stochastic phenomena such as stochastic [62–64] and coherence [65–71] resonances, noise-induced chimera states in networks of nonlocally-coupled excitable oscillators [72] and noise-induced patterns in excitable media [73]. Thus, delayed feedback is a powerful tool for achieving a wide range of operating regimes, enhancing amplitude-frequency characteristics, and controlling the dynamics of nonlinear systems [74–76].

In this section, two approaches to develop electronic delay lines are discussed. Both of them are used by the author in the frameworks of experimental studies of delayed-feedback oscillators.

4.1.1 Microcontroller-Based Delay Line

Consider the delay line depicted in Fig. 4.1. The central element of the circuit is the microcontroller ATMEGA16 connected to analog-to-digital converter MCP3201 using the SPI serial interface. The circuit's input and output signals are $x_{\text{in}}(t)$ and $x_{\text{out}} = x(t-\tau) \in [-2.5 : 2.5]$ V where τ is a delay time varying in the range from zero to 1.78 ms. Concrete values of τ linearly depend on voltage signal V_τ (is acquired by an internal ADC converter of the microcontroller) as $\tau = 3.56 \times 10^{-4} \times V_\tau$ and take 256 values equidistantly distributed in the range [0:1.78] ms at step $\Delta\tau \approx 6.9$ ms. It is important to note that zero voltage $V_\tau = 0$ corresponds to extremely small delay assumed to be zero. After the delayed signal is prepared, it is converted back into the analog form by a resistor ladder. In particular, R-2R ladder configuration in Fig. 4.1 transforms the digital signal $x_d(t-\tau)$ (8-bit) into the analog form $x_a(t-\tau)$ in [0:5] V. Finally, the circuit's output signal is obtained as $x_{\text{out}} = x_a(t-\tau) - 2.5$.

The program for ATMEGA16 is written in the C++ language adopted for AVR-family microcontrollers (the used software is AVR Studio) and represents a circular-buffer-based algorithm implementation. The author is grateful to Prof.

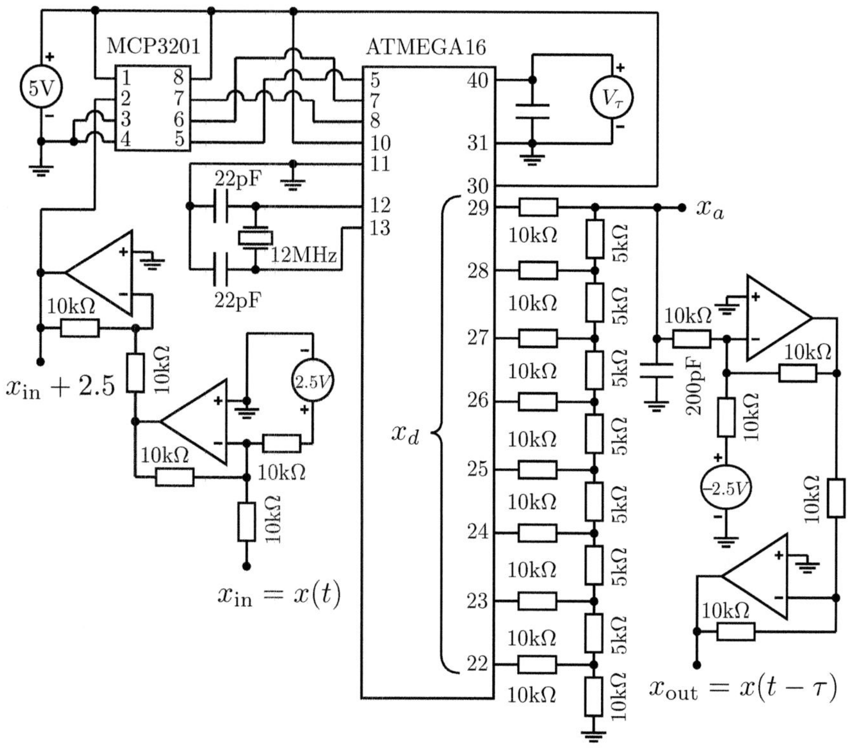

Fig. 4.1 Microcontroller-based implementation of the delay line

Anatoly Karavaev for helpful discussions and detailed explanations of the algorithm. In addition, the author thanks Dr. Boris Fayfel for lessons in C++ and Assembler stimulating the progress of the delay line implementation. Support by colleagues and friends made it possible to realize the electronic delayed feedback applied for experimental control of coherence resonance in non-excitable [71] and excitable [77] systems.

The program is listed below with short comments. The corresponding fuse bit combination used to configure the microcontroller is the following: CKSEL0=1, CKSEL1=1, CKSEL2=1, CKSEL3=1, SUT0=0, SUT1=0, BODEN=0, BODLEVEL=1, BOOTRST=1, BOOTSZ0=1, BOOTSZ1=1, EESAVE=1, CKOPT=0, JTAGEN=1, OCDEN=1.

The developed program:

```
// Program for the circular buffer algorithm implementation
// by Vladimir Semenov

// Import the library for programming ATMEGA16
#include <MEGA16.h>
// The microcontroller's clock frequency is 12 MHz
```

```
//This value is used as the variable fck in the comments
#define F_CPU 12000000

// Renaming variables and bits:

//Bits of the register ADCSRA being responsible for operation
// of the internal ADC acquiring signal V_tau.
#define ADEN 7
#define ADSC 6
#define ADATE 5
#define ADIF 4
#define ADIE 3
#define ADPS2 2
#define ADPS1 1
#define ADPS0 0

// Bits of the register ADMUX controlling the operation of
// the internal ADC
#define REFS1 7
#define REFS0 6
#define ADLAR 5
#define MUX4 4
#define MUX3 3
#define MUX2 2
#define MUX1 1
#define MUX0 0

// Bits of the register SFIOR controlling the operation of
// the internal ADC
#define ADTS2 7
#define ADTS1 6
#define ADTS0 5

// Pins of port B used for the SPI-interface implementation
#define DDR_SPI DDRB
#define DD_SS 4
#define DD_MOSI 5
#define DD_MISO 6
#define DD_SCK 7

// Bits of the register SPCR, being responsible for SPI
#define SPIE 7
#define SPE 6
#define DORD 5
#define MSTR 4
```

```
#define CPOL 3
#define CPHA 2
#define SPR1 1
#define SPR0 0

// Bits of the register SPSR used for SPI functioning
#define SPIF 7
#define SPI2X 0

// Procedures and functions

//Function for configuring the internal ADC
void internal_ADC_int()
{
   ADCSRA = (1<<ADEN) | (1<<ADSC) | (1<<ADATE);
   ADCSRA = (1<<ADPS2) | (1<<ADPS1) | (1<<ADPS0);
   // Enable using ADC, initialize the conversion in the
   // Free Running mode (continuous regime)
   // The ADC's frequency division factor is 128
   ADMUX = (1<<REFS0) | (1<<ADLAR);
   //The ADC's input is pin ADC0 (PA-0), AVCC with external
   //capacitor at AREF pin, prescaller is not used,
   //the results are left justified
   //Register SFIOR is not changed
}

//Function for initialization of the SPI interface
void spi_master_init()
{
   DDR_SPI = (1<<DD_MOSI) | (1<<DD_SCK) | (1<< DD_SS);
   // Configure the used pins of port B as outputs
   SPCR = (1<<SPE) | (1<<MSTR) | (0<<SPR1) | (0<<SPR0);
   SPSR = (1<<SPI2X);
   // Enable SPI, microcontroller acts as a master SPI device
   //  SPI Clock Rate is fck/2
}

//Byte transmission function
void SPI_MasterTransmit(char cData)
{
   SPDR = cData; //Start the transmission
   while(!(SPSR & (1<<SPIF)));// Wait the process end
}

//Acquire the instantaneous value of the input voltage signal
```

```
unsigned char ReceiveADC()
{
   unsigned char adcMSB=0;
   unsigned char adcLSB=0;
   unsigned char adc=0;
   PORTB &= ~(1<<DD_SS);
   SPI_MasterTransmit(0x00);
   adcMSB=SPDR;
   SPI_MasterTransmit(0x00);
   adcLSB=SPDR;
   PORTB |= (1<<DD_SS);

   // Merge the data using bite operation
   adc = adcMSB;
   adc<<=3;
   adcLSB>>=5;
   adc |= adcLSB;
   return adc;
}

// The main function of the program

void main (void) {
   //Initialization of an array containing
   //10 elements of 8-bite length
   char x[256];
   char n=0;
   int i;
   DDRC =0b11111111; // Configure port C as the output
   internal_ADC_int();
   spi_master_init();
   for (i=0;i<128;i++)
   {
      x[i]=ReceiveADC(); //Fill the array by initial values
   }
   // Start the process of delayed signal generation until
   // the microcontroller is switched off
   while(1)
   {
   // Acquire the value V_tau which determines the delay time
   // The internal ADC is used
   if (ADCSRA&(1<<4))
   {
      n=ADCH>>1;
      ADCSRA |=(1<<4);
```

```
    }

    // Fill the circular buffer and transmit it to the output
    // pins (only 8 bits of 10 are used, see PORT C)
    // The external ADC is used for data acquisition
    for (i=0;i<n;i++)
      {
        if (i==0)
        {
          PORTC = x[0];
          x[n-1]= ReceiveADC();
        }
        else
        {
          PORTC = x[i];
          x[i-1]= ReceiveADC();
        }
      }
      }
}
// The end
```

The presented microcontroller-based implementation of the delay line is not ideal and works correctly when the main frequency of the input signal does not exceed 5-10 KHz. To increase maximal time delay, one must use an external memory block. In addition, one can modify the program and the circuit for higher output signal accuracy by increasing the numbers of bites in the digital signal x_d.

4.1.2 PC-Based Delay Line

The second approach for the development of the delay line involves a personal computer (PC). In such a case, the used computer must be equipped by special boards to convert the analog input signal into a digital form, produce the delayed signal and then generate it as an analog output signal. The combination of software and hardware usually used by the author is depicted in Figs. 4.2 and 4.3. Here, the PC is equipped by the acquisition board NI PCI-6251 by National Instruments. The board is used to acquire the input analog signal $x(t)$ and convert it into a digital form for further processing by PC. Then the delayed signal $x(t-\tau)$ is generated as an analog output signal. To realize such scheme, the program was developed on the platform LabVIEW. The program consists of two parts: a front panel and a block diagram (Fig. 4.3). Here, the front panel is not depicted since it does not include any significant elements except of a button to stop the program. The block diagram contains two DAQ-assistant instruments. The first block 'DAQ Assistant'

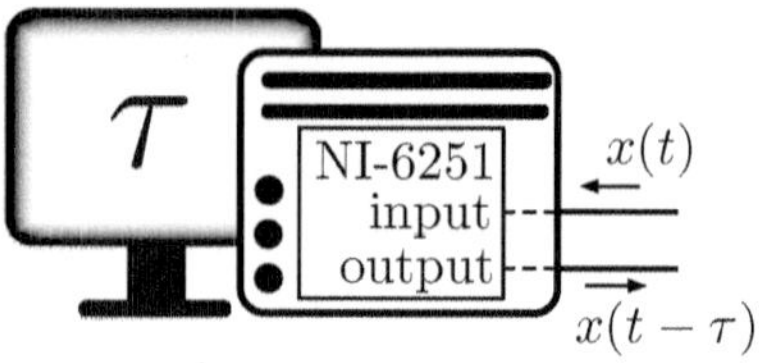

Fig. 4.2 Schematic illustration of the PC-based delay line implementation

is used to configure the input data acquisition (it contains a set of parameters such as a regime of acquisition, a list of used input channels, a sampling rate, an amplitude range, etc.). The second block 'DAQ Assistant 2' is responsible for analog output signal generation and allows to set the corresponding parameters. It is important to note that the output signal generation begins 5 seconds after the program is launched to provide for the initialization of all the processes. Inside the program, the delayed signal is produced by block 'Delay Values' situated between the DAQ-assistant blocks. It contains a controllable parameter being responsible for delay time. The minimal and maximal possible delay times depend on the personal computer characteristics. The intrinsic peculiarity of such delay line is large values of minimal delay time which cannot be assumed to be zero (the usual values are 3–10 ms).

The presented delay line was successfully applied by the author to implement delayed-feedback oscillators considered as a spatially-extended system. In particular, a single-signal delay line was used to realize dissipative solitons in a single delayed-feedback oscillator [21] (see Sect. 5.6). In addition, a similarly organized delay line proposed for simultaneous generation of two delayed signals for implementing a system of two delayed-feedback oscillators with identical delay times [78].

4.2 Generation of Noise

Noise was considered solely as a destructive factor for a long time. It was thought that the presence of noise only impairs and destroys any dynamical system functioning. However, the situation changed radically after a number of fundamental noise-induced effects was highlighted in nonlinear dynamical systems. In particular, if has been found that the influence of noise on nonlinear oscillators can lead to inducing new, more regular regimes, and to spatial pattern formation. A manifold of nontrivial noise-induced effects includes stochastic resonance [79, 80], coherence resonance [81, 82], noise-induced transitions [83, 84] and stochastic bifurcations [85–88], noise-induced synchronization [89–91], noise-induced chaos [92, 93].

To experimentally study stochastic effects, one can apply commercial broadband Gaussian noise generators and, if necessary, process their output signals to realize coloured noise or other kinds of the stochastic impact. The second approach is noise generation by means of personal computers equipped by electronic boards and by

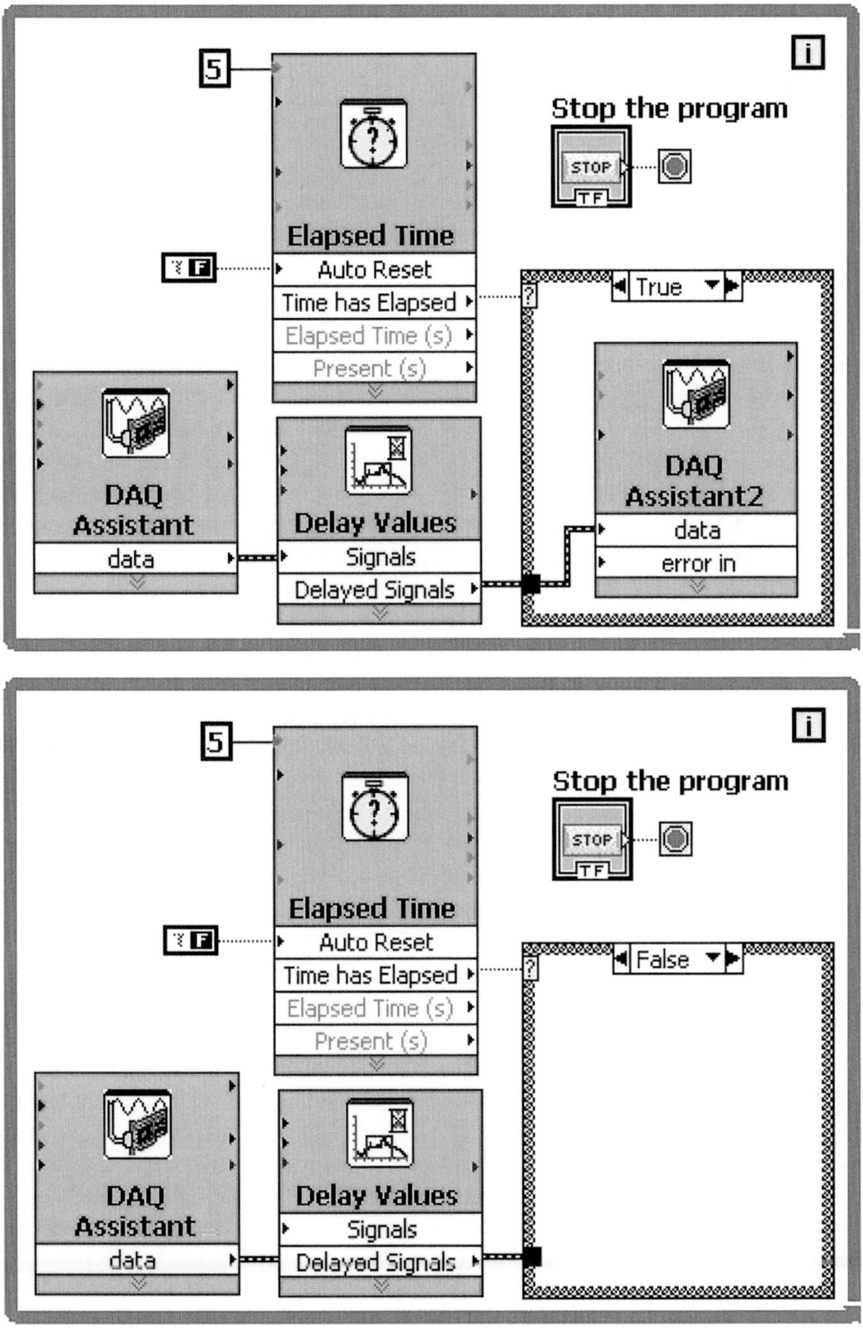

Fig. 4.3 Block diagram of the LabVIEW program developed for the PC-based delay line implementation. The program contains a block acting as the 'if' operator. The upper panel illustrate the case of obeying the condition of 5 second elapsing, whereas the lower panel corresponds to failure to comply with the condition

using special programs (for instance, developed on platform LabVIEW). The third way is much cheaper and simpler and involves 'home-made' noise generators (for instance, see a brief discussion of different implementations in review [94]). All the mentioned strategies are discussed below.

4.2.1 Broadband Gaussian Noise

In many cases, experiments on stochastic dynamical systems are carried out to replicate oscillations induced by white Gaussian noise and obtained earlier in numerical simulations. This kind of noise represents an idealization and is characterised by zero correlation time resulting in flat power spectrum up to infinite frequency. In practice, noise generated by real physical systems has a finite correlation time and a power spectrum being flat in a some finite frequency range (for this reason, the term 'white' is changed to 'broadband' in the current section). However, experimental noise can be considered as the white one when the noise correlation time is much smaller than all characteristic times of a noise-driven system (response times, oscillation period, relaxation time, etc.).

As already mentioned above, the simplest, the best, but not the cheapest way to generate broadband Gaussian noise for further study of noise-induced effects is using commercial devices developed and manufactured by famous companies. For instance, one of such devices is generator Agilent 33250A with function of noise generating. Its output noise signal (one of the time realizations is depicted in Fig. 4.4a) is characterised by the distribution of the probability density function (see Fig. 4.4b) being very close to Gaussian. The corresponding power spectrum is almost flat in a range up to 50 MHz. The shape of the power spectrum of the stochastic process in Fig. 4.4a is depicted in Fig. 4.4c and indicates that the generator output signal can be considered as white noise at least in the range [0–1] MHz. This range is pretty enough to study noise-induced effects on examples of electronic circuits which main oscillation frequency does not exceed tens of kHz and consider them to be driven by white noise.

The second strategy for generation of broadband Gaussian noise involves programming approach. As an example, such program developed on the base of LabVIEW platform is illustrated in Fig. 4.5. It contains an instrument for continuous simulation of white Gaussian noise as a virtual signal. Then this signal is generated by NI PCI-6251 as an analog output. In a virtual form, the simulated signal is assumed to be white noise. After the final generation of the signal as an output voltage, it remains to be Gaussian noise (see the time realization and the corresponding PDF in Fig. 4.6a, b). However, the power spectrum of the physical signal allows to assume the noise to be white in the frequency range [0:200] kHz (see Fig. 4.6c). If this signal affects an electronic oscillator which main oscillation frequency does not exceed 7–10 kHz, the oscillator can be considered as a dynamical system subject to white Gaussian noise. Nevertheless, it should be noted that the power spectrum fully corresponds to coloured noise in the range [0:1] MHz.

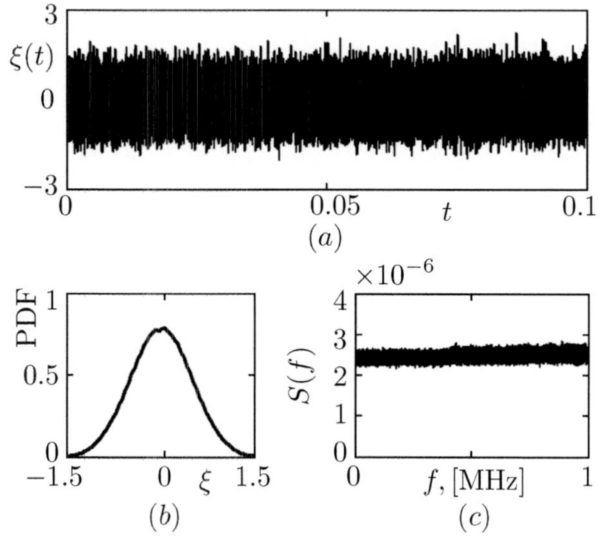

Fig. 4.4 Time realization (panel (**a**)), distribution of the probability density function (panel (**b**)) and power spectrum (panel (**c**)) of noise produced by Agilent 33250A. The corresponding noise standard deviation is $\sigma = 0.5$

Fig. 4.5 Block diagram of the LabVIEW program developed for the broadband Gaussian noise generation

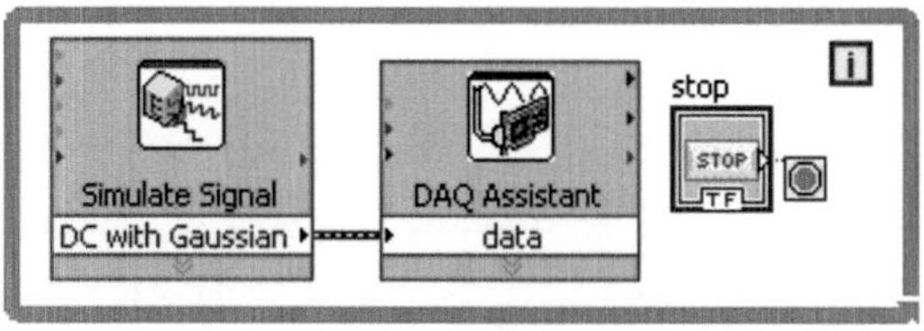

An advantage of the noise generation based on a PC equipped by NI acquisition boards is the ability to simultaneously generate a number of statistically independent noise sources which is determined by a number of analog outputs (for instance, board NI PCI-6251 has two analog outputs).

If one deals with a rough noise-induced effect and the correspondence with a mathematical model is not principal, the implementation of the effect by means of electronics can be realized by using hand-made noise sources. One example of such circuits is depicted in Fig. 4.7a. The operation principle of the circuit is based on the properties of the Zener diodes (see diodes BZX85-C6V8 in Fig. 4.7a). In particular, the voltage applied across the diodes is close to the breakdown voltage value and provides for generation of a signal with well pronounced fluctuations (see the current-voltage characteristic in Fig. 4.7b). For normal circuit operation, the applied voltage must be well-stabilized. To provide this, fixed voltage regulator LM7812 is used. The circuit contains two stochastic signals produced by the Zener diodes (signals V1 and V2 in Fig. 4.7a) which are input signals for difference amplifier. The presence of the difference amplifier processing two identical but

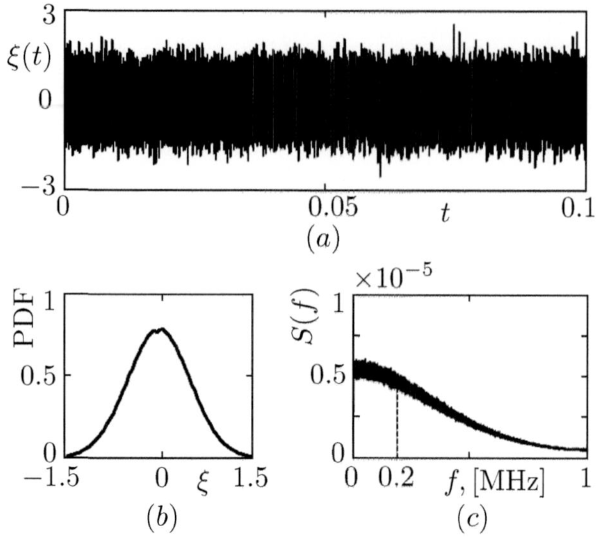

Fig. 4.6 Time realization (panel (**a**)), distribution of the probability density function (panel (**b**)) and power spectrum (panel (**c**)) of Gaussian noise (standard deviation is $\sigma = 0.5$) produced by PC equipped by board NI PCI-6251 (see the corresponding program in Fig. 4.5)

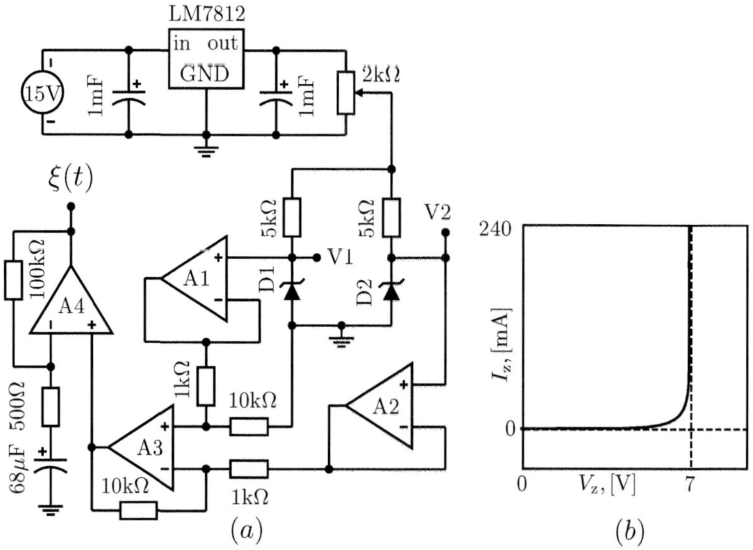

Fig. 4.7 (**a**) Noise source based on the Zener diode. Diodes D1 and D2 are BZX85-C6V8. Operational amplifiers are LF412CN (A1 and A2) and LM6172 (A3 and A4); (**b**) Current-voltage characteristic of Zener Diode BZX85-C6V8 (see the corresponding datasheet for more detail). The breakdown voltage is close to 7 Volts

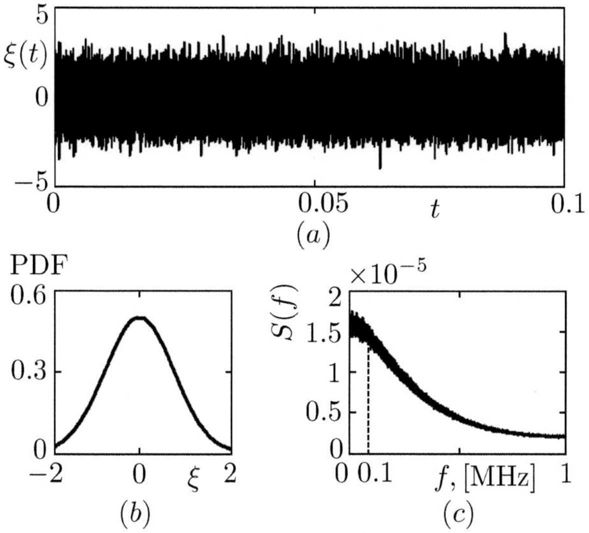

Fig. 4.8 Time realization (panel (**a**)), distribution of the probability density function (panel (**b**)) and power spectrum (panel (**c**)) of Gaussian noise (measured standard deviation is $\sigma = 0.8$) produced by the circuit in Fig. 4.7a

independent stochastic signals allows to obtain noise with symmetric distribution of the probability density function (see the PDF in Fig. 4.8b). Finally, the difference amplifier output signal is gained by the non-inverting amplifier to obtain a high-amplitude noise $\xi(t)$.

4.2.2 Coloured Noise (Ornstein-Uhlenbeck Process)

The assumption to use uncorrelated (also called δ-correlated) random perturbation model is good and useful if the time scale of noise (associated with the properties of the correlation function) is much shorter than the characteristic time scale of a stochastically forced dynamical system. This assumption significantly simplifies the model and simulation. However, this idealization cannot be exactly realized in real physical systems. In order to understand the role and significance of the noise finite correlation, it is necessary to consider non-white noise model also called coloured noise.

The properties of coloured noise take into consideration a new characteristic, a correlation time which can be used as an independent control parameter to determine qualitative nature of the dynamics, as well as to tune its quantitative indicators [95]. For instance, the intrinsic peculiarities of coloured noise are used to control stochastic [96, 97] and coherence [69–71, 77] resonances, stochastic bifurcations [98] and noise-induced transitions [99].

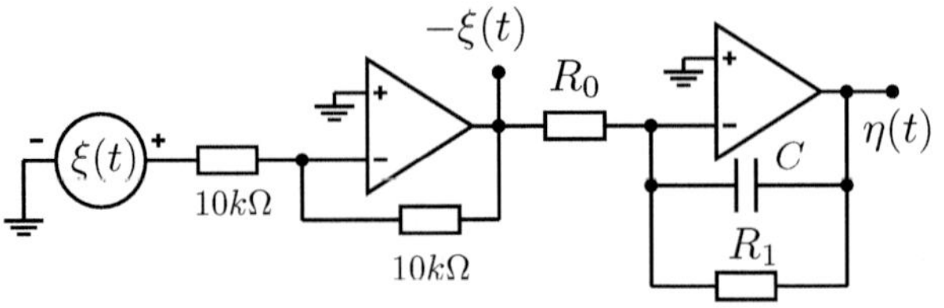

Fig. 4.9 Electronic circuit implementation of the coloured noise source as a realization of the first-order Ornstein-Uhlenbeck process

Technically, the coloured noise is a random process that has a variable power in a band of a given bandwidth. Such stochastic signals can be realized by using different filters. If it is principally important to implement the coloured noise source as a rigorous model of a stationary Gauss-Markov process, one can realize the first-order Ornstein-Uhlenbeck model by means of analog modelling. One of the possible circuits is depicted in Fig. 4.9. The circuit equation is $R_0 C\frac{d\eta}{dt} = -\frac{R_0}{R_1}\eta + \xi(t)$ (here $\xi(t)$ is a source of noise assumed to be white and Gaussian), which can be rewritten in the dimensionless form $\tau_c\frac{d\eta}{dt} = -\eta + \sqrt{2D\tau_c}n(t)$ by introducing new time scales depending on the properties of a dynamical system forced by stochastic signal $\eta(t)$. Considering the dimensionless form, one deal with two controllable parameters: τ_c is the correlation time of noise, D is the noise intensity. The equation contains an additive source of white Gaussian noise $\sqrt{2D\tau_c}n(t)$ which means zero mean value, $< n(t) >= 0$, and the correlation function described by the delta function, $< n(t)n(t+\tau) >= 2D\tau_c\delta(\tau)$.

The circuit in Fig. 4.9 can be simplified by reduction of the inverting amplifier. Since the noise source $\xi(t)$ is assumed to be Gaussian characterised by symmetric distribution of the probability density, multiplying $\xi(t)$ by -1 has no principal sense.

4.2.3 Levy Noise

Lévy noise is a class of stable non-Gaussian noise that exhibits long heavy tails of its distribution associated with large, potentially infinite, jumps. Lévy noise is often used to model stochastic systems where the noise-induced dynamics is characterised by abrupt changes. One can observe such effects in lasers [100], cardiac rhythms [101], molecular motors [102], quantum dots [103], financial [104, 105] and social [106] systems. Moreover, stochastic processes with a Lévy distribution can model the dynamics of real biological neurons more accurately as compared to Gaussian noise [107, 108].

In the context of noise-induced phenomena, Lévy noise is known to provide for inducing and suppressing travelling waves [109] and controlling characteristics of noise-induced oscillations in the regime of stochastic resonance [110–112], self-induced stochastic resonance [113], coherence resonance in non-excitable [112] and excitable [114] systems.

Generally, Lévy noise is defined as the formal derivative of the Lévy stable motion and is characterized by four parameters: stability index $\alpha \in (0:2]$, skewness (asymmetry) parameter $\beta \in [-1:1]$, mean value μ (μ is set to be zero for the strictly stable distributions [115]) and scale parameter σ. Usually, the noise intensity is introduced as $D = \sigma^{\alpha}$.

If the noise signal $\xi(t)$ obeys to Lévy distribution $L_{\alpha,\beta}(\xi, \sigma, \mu)$, its characteristic function $\phi(k) = \int\limits_{-\infty}^{+\infty} \exp(ikx) L_{\alpha,\beta}(\xi, \sigma, \mu)dx$ takes the form [110, 115, 116]:

$$\phi(k) = \begin{cases} \exp\left[-\sigma^{\alpha}|k|^{\alpha}\left(1 - i\beta sgn(k)\tan\frac{\pi\alpha}{2}\right)\right], & \text{for } \alpha \neq 1, \\ \exp\left[-\sigma|k|\left(1 + i\beta\frac{2}{\pi}sgn(k)\ln|k|\right)\right], & \text{for } \alpha = 1. \end{cases} \tag{4.1}$$

To generate random sequence ξ corresponding to characteristic function (4.1), one can apply the Janicki-Weron algorithm [115, 117]:

$$\begin{aligned} &\text{For } \alpha \neq 1: \\ &\xi = \sigma S_{\alpha,\beta} \times \frac{\sin(\alpha(V + B_{\alpha,\beta}))}{(\cos(V))^{1/\alpha}} \times \left(\frac{\cos(V - \alpha(V + B_{\alpha,\beta}))}{W}\right)^{\frac{1-\alpha}{\alpha}}, \\ &\text{For } \alpha = 1: \\ &\xi = \frac{2\sigma}{\pi}\left[\left(\frac{\pi}{2} + \beta V\right)\tan(V) - \beta\ln\left(\frac{\frac{\pi}{2}W\cos(V)}{\frac{\pi}{2} + \beta V}\right)\right], \end{aligned} \tag{4.2}$$

where $B_{\alpha,\beta} = \left(\operatorname{atan}\left(\beta\tan\left(\frac{\pi\alpha}{2}\right)\right)\right)/\alpha$, $S_{\alpha,\beta} = \left(1 + \beta^2\tan^2\left(\frac{\pi\alpha}{2}\right)\right)^{1/2\alpha}$, V is a random variable uniformly distributed on $\left(-\frac{\pi}{2} : \frac{\pi}{2}\right)$, W is an exponential random variable with mean 1 (variables W and V are independent). In case $\alpha = 2$, the distribution of the probability density function takes the Gaussian form with zero mean value and the variance being equal to $2\sigma^2$. If $\alpha < 2$, the distribution is non-Gaussian and the variance is infinite. Numerical simulations of dynamical systems forced by Lévy noise are carried out by the integration of the model equations with taking into account the fact that numerical modelling equations including α-stable stochastic process with finite time step implies the normalization of the noise term by $\Delta t^{1/\alpha}$ [118, 119].

A source of Lévy noise can be implemented using a personal computer complemented by an electronic board including an analog output (for instance, National Instruments NI-PCI 6251). To create a source of Lévy noise, an algorithm of random number generation (4.2) can be realized on the Labview platform. One

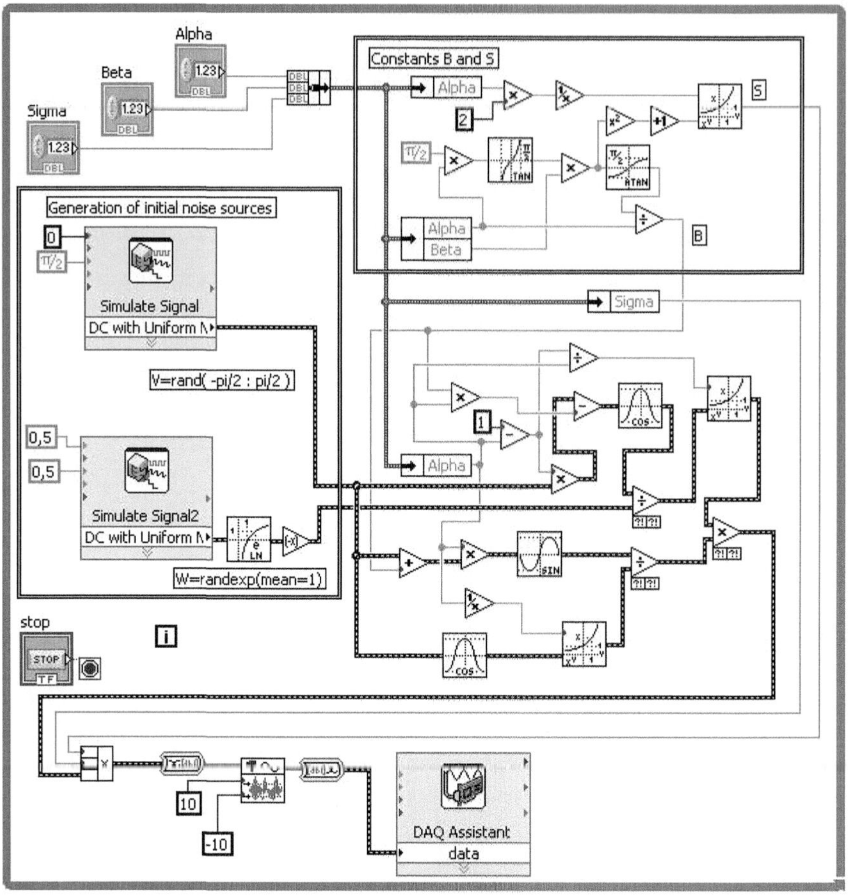

Fig. 4.10 Program developed on the Labview platform for electronic implementation of a PC-based Lévy noise source. In the presented version, the noise parameter μ is always equal to zero

such solutions is demonstrated in Fig. 4.10. The Labview program includes a block being responsible for generating random numbers ξ with high frequency (up to 10^6 samples per second) as an output analog signal (voltage) of the NI-board. The noise signal corresponding to $\alpha = 2$, $\beta = \mu = 0$ generated by using Labview program in Fig. 4.10 is depicted in Fig. 4.11 as well as its power spectrum and the distribution of the probability density function. As can be seen in Fig. 4.11, the signal exhibits the properties of Gaussian noise at $\alpha = 2$. However, the noise signal differs from the Gaussian noise form when decreasing α. In such a case, the noise signal is characterised by the appearance of sharp jumps up and down (Fig. 4.12a). When $\beta \neq 0$, either jumps up or down are more pronounced (Fig. 4.12b, c). In general, the properties of the experimentally generated signal are very close to the peculiarities

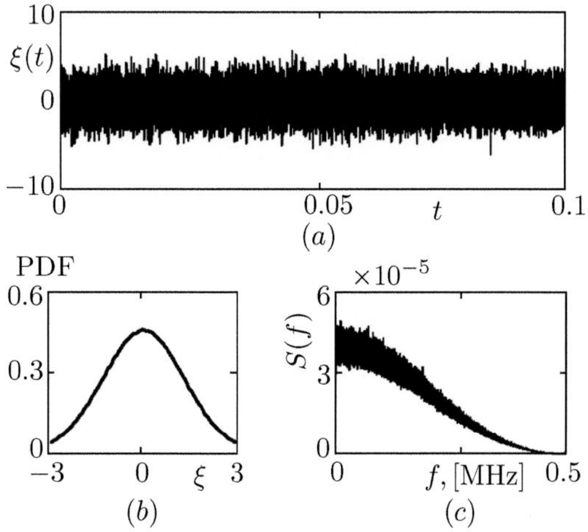

Fig. 4.11 Time realization (panel (**a**)), distribution of the probability density function (panel (**b**)) and power spectrum (panel (**c**)) of the noise signal generated using the program in Fig. 4.10. Parameters of noise are $\alpha = 2$, $\beta = 0$, $\mu = 0$, $\sigma = 1$

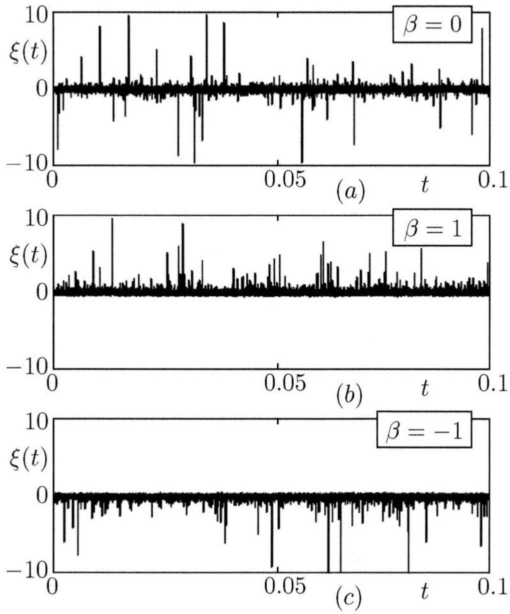

Fig. 4.12 Noise signals generated using the program in Fig. 4.10. Parameters of noise are $\alpha = 1.8$, $\mu = 0$, $\sigma = 0.1$, $\beta = 0$ (panel (**a**)), $\beta = 1$ (panel (**b**)), $\beta = -1$ (panel (**c**))

of the Lévy noise model. The only difference is associated with properties of the acquisition board: the amplitude of signal produced by NI-PCI 6251 is limited by the range [−10:10] V. In certain cases, this limitation can be avoided. If the produced noise is an input signal of an analog integrator, one can decrease the corresponding input resistance to amplify the resulting term in an electronic model equation. Such approach was applied in paper [114] on example of Lévy noise-induced coherence resonance.

References

1. T. Erneux, J. Javaloyes, M. Wolfrum, S. Yanchuk, Chaos Interdiscip. J. Nonlinear Sci. **27**(11), 114201 (2017)
2. S. Crook, G. Ermentrout, M. Vanier, J. Bower, J. Comput. Neurosci. **4**(2), 161 (1997)
3. G. Orosz, J. Moehlis, R.M. Murray, Philos. Trans. R. Soc. A Math. Phys. Eng. Sci. **368**(1911), 439 (2010)
4. Y. Kuang (ed), *Delay Differential Equations with Applications in Population Dynamics*. Mathematics in Science and Engineering, vol. 191 (Academic Press, Cambridge, 1993)
5. S. Pal, M. Hossain, P. Panday, N.C. Pati, N. Pal, J. Chattopadhyay, Chaos Interdiscip. J. Nonlinear Sci. **30**(8), 083124 (2020)
6. B. Rosin, M. Slovik, R. Mitelman, M. Rivlin-Etzion, S.N. Haber, Z. Israel, E. Vaadia, H. Bergman, Neuron **72**(2), 370 (2011)
7. S. Little, A. Pogosyan, S. Neal, B. Zavala, L. Zrinzo, M. Hariz, T. Foltynie, P. Limousin, K. Ashkan, J. FitzGerald, A.L. Green, T.Z. Aziz, P. Brown, Ann. Neurol. **74**(3), 449 (2013)
8. O.V. Popovych, P.A. Tass, Front. Physiol. **9**, 46 (2018)
9. M. Rosenblum, A. Pikovsky, Phys. Rev. E **70**(4), 041904 (2004)
10. J. Modolo, A. Legros, A.W. Thomas, A. Beuter, Interface Focus **1**(1), 61 (2010)
11. B.B. Ferreira, A.S. de Paula, M.A. Savi, Chaos Solitons Fract. **44**(8), 587 (2011)
12. L.S. Young, S. Ruschel, S. Yanchuk, T. Pereira, Sci. Rep. **9**(1), 3505 (2019)
13. L. Bauer, J. Bassett, P. Hövel, Y.N. Kyrychko, K.B. Blyuss, Chaos Interdiscip. J. Nonlinear Sci. **27**(11), 114317 (2017)
14. M. Schell, J. Ross, J. Chem. Phys. **85**(11), 6489 (1986)
15. R.A. van Santen, J. Chem. Phys. **57**(12), 5418 (1972)
16. S. Tang, J. Liu, IEEE J. Quant. Electron. **37**(3), 329 (2001)
17. F. Yi Lin, J.M. Liu, IEEE J. Quant. Electron. **39**(4), 562 (2003)
18. D. Brunner, B. Penkovsky, R. Levchenko, E. Schöll, L. Larger, Y. Maistrenko, Chaos Interdiscip. J. Nonlinear Sci. **28**(10), 103106 (2018)
19. Y.K. Chembo, D. Brunner, M. Jacquot, L. Larger, Rev. Mod. Phys. **91**(3), 035006 (2019)
20. T. Banerjee, D. Biswas, *Time-Delayed Chaotic Dynamical Systems: From Theory to Electronic Experiment*. SpringerBriefs in Applied Sciences and Technology, vol. XI (Springer, Berlin, 2018)
21. V.V. Semenov, Y.L. Maistrenko, Chaos Interdiscip. J. Nonlinear Sci. **28**(10), 101103 (2018)
22. R. Sipahi, S.I. Niculescu, IFAC Proc. Vol. **40**(23), 111 (2007)
23. A. Keane, B. Krauskopf, C.M. Postlethwaite, Chaos Interdiscip. J. Nonlinear Sci. **27**(11), 114309 (2017)
24. M. Gassel, E. Glatt, F. Kaiser, Fluct. Noise Lett. **07**(03), L225 (2007)
25. V.K. Jirsa, Cognit. Neurodyn. **2**(1), 29 (2007)
26. V.K. Jirsa, Philos. Trans. R. Soc. A Math. Phys. Eng. Sci. **367**(1891), 1131 (2009)
27. Z. Wang, S.A. Campbell, Chaos Interdiscip. J. Nonlinear Sci. **27**(11), 114316 (2017)
28. Y. Han, S. Xiang, Z. Ren, C. Fu, A. Wen, Y. Hao, Photon. Res. **9**(4), B119 (2021)
29. X. Wang, X. Lin, X. Dang, Front. Neurosci. **13**, 252 (2019)
30. G. Tanaka, T. Yamane, J.B. Héroux, R. Nakane, N. Kanazawa, S. Takeda, H. Numata, D. Nakano, A. Hirose, Neural Netw. **115**, 100 (2019)
31. D. Brunner, M.C. Soriano, G.V. der Sande (eds), *Photonic Reservoir Computing* (De Gruyter, Berlin, 2019)
32. R. Martinenghi, S. Rybalko, M. Jacquot, Y.K. Chembo, L. Larger, Phys. Rev. Lett. **108**(24), 244101 (2012). https://doi.org/10.1103/physrevlett.108.244101
33. D. Brunner, B. Penkovsky, B.A. Marquez, M. Jacquot, I. Fischer, L. Larger, J. Appl. Phys. **124**(15), 152004 (2018)
34. L. Larger, A. Baylón-Fuentes, R. Martinenghi, V.S. Udaltsov, Y.K. Chembo, M. Jacquot, Phys. Rev. X **7**(1), 011015 (2017)

35. G.V. der Sande, D. Brunner, M.C. Soriano, Nanophotonics **6**(3), 561 (2017)
36. B. Penkovsky, X. Porte, M. Jacquot, L. Larger, D. Brunner, Phys. Rev. Lett. **123**(5), 054101 (2019)
37. M. Goldmann, F. Köster, K. Lüdge, S. Yanchuk, Chaos Interdiscip. J. Nonlinear Sci. **30**(9), 093124 (2020)
38. F. Stelzer, A. Röhm, R. Vicente, I. Fischer, S. Yanchuk, Nat. Commun. **12**(1), 5164 (2021)
39. F. Böhm, G. Verschaffelt, G.V. der Sande, Nat. Commun. **10**(1) (2019)
40. E. Schöll, G. Hiller, P. Hövel, M.A. Dahlem, Philos. Trans. R. Soc. A Math. Phys. Eng. Sci. **367**(1891), 1079 (2009)
41. N. Burić, D. Todorović, Phys. Rev. E **67**(6), 066222 (2003)
42. M.A. Dahlem, G. Hiller, A. Panchuk, E. Schöll, Int. J. Bifur. Chaos **19**(02), 745 (2009)
43. O. Vallès-Codina, R. Möbius, S. Rüdiger, L. Schimansky-Geier, Phys. Rev. E **83**(3), 036209 (2011)
44. V. Semenov, A. Bukh, N. Semenova, Chaos Solitons Fract. **172**, 113524 (2023)
45. M. Schanz, A. Pelster, SIAM J. Appl. Dyn. Syst. **2**(3), 277 (2003)
46. N.B. Janson, C.J. Marsden, Chaos Interdiscip. J. Nonlinear Sci. **31**(11), 113126 (2021)
47. J. Hizanidis, R. Aust, E. Schöll, Int. J. Bifur. Chaos **18**(06), 1759 (2008)
48. M.M. Asl, A. Valizadeh, P.A. Tass, Sci. Rep. **8**(1), 12068 (2018)
49. K. Pyragas, Phys. Lett. A **170**(6), 421 (1992)
50. P. Hövel, E. Schöll, Phys. Rev. E **72**(4), 046203 (2005)
51. J.H. Yang, X.B. Liu, J. Phys. A Math. Theor. **43**(12), 122001 (2010)
52. D. Hu, J. Yang, X. Liu, Commun. Nonlinear Sci. Numer. Simul. **17**(2), 1031 (2012)
53. J. Farmer, Phys. D **4**(3), 366 (1982)
54. M. Krupa, J.D. Touboul, J. Nonlinear Sci. **26**(1), 43 (2016)
55. O. Beck, A. Amann, E. Schöll, J.E.S. Socolar, W. Just, Phys. Rev. E **66**(1), 016213 (2002)
56. J. Unkelbach, A. Amann, W. Just, E. Schöll, Phys. Rev. E **68**(2), 026204 (2003)
57. A.C. Martí, C. Masoller, Phys. Rev. E **67**(5), 056219 (2003)
58. V.V. Klinshov, V.I. Nekorkin, Phys. Uspekhi **56**(12), 1217 (2013)
59. C. Nathe, K. Huang, M. Lodi, M. Storace, F. Sorrentino, Chaos Interdiscip. J. Nonlinear Sci. **30**(12), 121105 (2020)
60. O.V. Popovych, C. Hauptmann, P.A. Tass, Phys. Rev. Lett. **94**(16), 164102 (2005)
61. O.V. Popovych, B. Lysyansky, M. Rosenblum, A. Pikovsky, P.A. Tass, PLOS ONE **12**(3), e0173363 (2017)
62. D. Wu, S. Zhu, Phys. Lett. A **372**(32), 5299 (2008)
63. D.C. Mei, L.C. Du, C.J. Wang, J. Stat. Phys. **137**(4), 625 (2009)
64. Z.L. Jia, Phys. Script. **81**(1), 015002 (2009)
65. N.B. Janson, A.G. Balanov, E. Schöll, Phys. Rev. Lett. **93**(1), 010601 (2004)
66. A. Balanov, N. Janson, E. Schöll, Phys. D Nonlinear Phenom. **199**(1–2), 1 (2004)
67. T. Prager, H.P. Lerch, L. Schimansky-Geier, E. Schöll, J. Phys. A Math. Theor. **40**(36), 11045 (2007)
68. R. Aust, P. Hövel, J. Hizanidis, E. Schöll, Eur. Phys. J. Special Top. **187**(1), 77 (2010). https://doi.org/10.1140/epjst/e2010-01272-5
69. S. Brandstetter, M.A. Dahlem, E. Schöll, Phil. Trans. R. Soc. A **368**(1911), 391 (2010)
70. P.M. Geffert, A. Zakharova, A. Vüllings, W. Just, E. Schöll, Eur. Phys. J. B **87**, 291 (2014)
71. V. Semenov, A. Feoktistov, T. Vadivasova, E. Schöll, A. Zakharova, Chaos **25**(3), 033111 (2015)
72. A. Zakharova, N. Semenova, V. Anishchenko, E. Schöll, Chaos Interdiscip. J. Nonlinear Sci. **27**(11), 114320 (2017)
73. A.G. Balanov, V. Beato, N.B. Janson, H. Engel, E. Schöll, Phys. Rev. E **74**(1), 016214 (2006)
74. E. Schöll, H.G. Schuster (eds), *Handbook of Chaos Control* (Wiley, Hoboken, 2007)
75. J.Q. Sun, Q. Ding, *Advances in Analysis and Control of Time-Delayed Dynamical Systems* (World Scientific, Singapore, 2013)

76. V. Flunkert, I. Fischer, E. Schöll, Philos. Trans. R. Soc. A Math. Phys. Eng. Sci. **371**(1999), 20120465 (2013)
77. V. Semenov, T. Vadivasova, E. Schöll, A. Zakharova, Izvestiya of Saratov University. Physics [in Russian] **15**(3), 43 (2015)
78. A. Zakharova, V. Semenov, Delayed-feedback oscillators replicate the dynamics of multiplex networks: wavefront propagation and stochastic resonance. Neural Netw. **183**, 106939 (2025). https://doi.org/10.1016/j.neunet.2024.106939
79. L. Gammaitoni, P. Hänggi, P. Jung, F. Marchesoni, Rev. Mod. Phys. **70**(1), 223 (1998)
80. V. Anishchenko, A. Neiman, F. Moss, L. Schimansky-Geier, Phys. Uspekhi **42**(1), 7 (1999)
81. A. Pikovsky, J. Kurths, Phys. Rev. Lett. **78**(5), 775 (1997)
82. B. Lindner, J. García-Ojalvo, A. Neiman, L. Schimansky-Geier, Phys. Rep. **392**, 321 (2004)
83. W. Horsthemke, R. Lefever, *Noise-Induced Transitions* (Springer, Berlin, 2006)
84. I. Bashkirtseva, L. Ryashko, H. Schurz, Chaos Solitons Fractals **39**(1), 72 (2009)
85. L. Arnold, *Random Dynamical Systems* (Springer, Berlin, 2003)
86. R. Lefever, J. Turner, Fluctuations and Sensitivity in Nonequilibrium Systems, ed. by W. Horsthemke (Springer, Berlin, 1984), pp. 143–149
87. R. Lefever, J. Turner, Phys. Rev. Lett. **56**(16), 1631 (1986)
88. L. Fronzoni, R. Mannella, P. McClintock, F. Moss, Phys. Rev. A **36**(2), 834 (1987)
89. A. Neiman, L. Schimansky-Geier, A. Cornell-Bell, F. Moss, Phys. Rev. Lett. **83**(23), 4896 (1999)
90. D. Goldobin, A. Pikovsky, Phys. Rev. E **71**(4), 045201(R) (2005)
91. B. Hu, C. Zhou, Phys. Rev. E. **61**(2), R1001 (2000)
92. L. Schimansky-Geier, H. Herzel, J. Stat. Phys. **70**(1), 141 (1993)
93. I. Bashkirtseva, E. Ekaterinchuk, L. Ryashko, Int. J. Bifur. Chaos **26**(4), 1650053 (2016)
94. D. Luchinsky, P. McClintock, M. Dykman, Rep. Prog. Phys. **61**(8), 889 (1998)
95. P. Hänggi, P. Jung, *Advances in Chemical Physics* (Wiley, Hoboken, 1994). Chap. Colored noise in dynamical systems
96. P. Hänggi, P. Jung, C. Zerbe, F. Moss, J. Stat. Phys. **70**(1), 25 (1993)
97. D. Nozaki, D. Mar, P. Grigg, J. Collins, Phys. Rev. Lett. **82**(11), 2402 (1999)
98. Y. Xu, R. Gu, H. Zhang, W. Xu, J. Duan, Phys. Rev. E **83**(5), 056215 (2011)
99. K. Kitahar, W. Horsthemke, R. Lefever, Phys. Lett. A **70**(5–6), 377 (1979)
100. E. Rocha, E. Santos, B. dos Santos, S. de Albuquerque, P. Pincheira, C. Argolo, A. Moura, Phys. Rev. A **101**(2), 023820 (2020)
101. C.K. Peng, J. Mietus, J.M. Hausdorff, S. Havlin, H.E. Stanley, A.L. Goldberger, Phys. Rev. Lett. **70**(9), 1343 (1993)
102. B. Lisowski, D. Valenti, B. Spagnolo, M. Bier, E. Gudowska-Nowak, Phys. Rev. E **91**(4), 042713 (2015)
103. D.S. Novikov, M. Drndic, L.S. Levitov, M.A. Kastner, M.V. Jarosz, M.G. Bawendi, Phys. Rev. B **72**(7), 075309 (2005)
104. R. Mantegna, H. Stanley, *An Introduction to Econophysics* (Cambridge University Press, Cambridge, 1999)
105. O. Barndorff-Nielsen, N. Shephard, J. Royal Stat. Soc. Ser. B **63**(2), 167 (2001)
106. M. Perc, Phys. Rev. E **75**(2), 022101 (2007)
107. S. Nurzaman, Y. Matsumoto, Y. Nakamura, K. Shirai, S. Koizumi, H. Ishiguro, PLOS ONE **6**(2), 1 (2011)
108. J. Wu, Y. Xu, J. Ma, PLOS ONE **12**(3), 1 (2017)
109. I.A. Korneev, I.R. Ramazanov, A.V. Slepnev, T.E. Vadivasova, V.V. Semenov, Travelling waves in an ensemble of excitable oscillators: the interplay of memristive coupling and noise. Chaos **34**(12), 123110 (2024). https://doi.org/10.1063/5.0215152
110. B. Dybiec, E. Gudowska-Nowak, Acta Phys. Pol. B **37**(5), 1479 (2006)
111. B. Dybiec, E. Gudowska-Nowak, J. Stat. Mech. **2009**(5), P05004 (2009)
112. R. Yonkeu, R. Yamapi, G. Filatrella, J. Kurths, Eur. Phys. J. B **93**(8), 144 (2020)
113. M. Yamakou, T. Tran, Nonlinear Dyn. **107**(3), 2847 (2022)
114. I. Korneev, A. Zakharova, V. Semenov, Chaos Solitons Fractals **184**, 115037 (2024)

115. A. Janicki, A. Weron, *Simulation and Chaotic Behavior of Alpha-Stable Stochastic Processes* (Marcel Dekker, New York, 1994)
116. B. Dybiec, E. Gudowska-Nowak, P. Hänggi, Phys. Rev. E **75**(2), 021109 (2007)
117. A. Weron, R. Weron, *Computer Simulation of Lévy* Math-*Stable Variables and Processes* (Springer, Berlin, 1995), pp. 379–392
118. Y. Xu, Y. Li, H. Zhang, X. Li, J. Kurths, Sci. Rep. **6**(1), 31505 (2016)
119. I. Pavlyukevich, B. Dybiec, A. Chechkin, I. Sokolov, Eur. Phys. J. Special Top. **191**(1), 223 (2010)

Chapter 5
Dynamical Systems and the Corresponding Electronic Models

In this chapter all the techniques presented before as theoretical concepts are discussed on examples of concrete dynamical systems and circuits. For certain models, circuits developed by means of classical electronic engineering approach and analog modelling are shown to demonstrate advantages and disadvantages of both concepts. In addition to the dynamical system description, the behaviour exhibited by the considered model is explained to provide an overview on fundamental deterministic and stochastic phenomena constituting a basis for nonlinear dynamics and theory of stochastic processes.

5.1 Bistability as the Coexistence of Two Stable Equilibria

5.1.1 Kramers Oscillator with Finite Dissipation

Bistable dynamics is typical for many natural systems in physics [1–6], chemistry [7, 8], biology [9–15], ecology [16, 17], geophysics [18–20]. The simplest kind of bistability occurs when a system possesses two stable equilibria in the phase space, separated by a saddle. Adding noise gives rise to random switchings between the deterministically stable states, resulting in a steady state probability density with two local maxima. The Kramers oscillator is a classical example of the stochastic bistable system describing Brownian motion in a double-well potential [2, 7, 21],

$$\frac{d^2x}{dt^2} = -\gamma\frac{dx}{dt} - \frac{dU(x)}{dx} + \xi(t), \tag{5.1}$$

where γ is the (constant) drag coefficient (also called the coefficient of dissipation), $U(x)$ is a potential function, and $\xi(t)$ is a noise source. The regime of bistability is achieved when the function $U(x)$ describes a double-well potential, $U(x) =$

© The Author(s), under exclusive license to Springer Nature Switzerland AG 2024

V. Semenov, *Electronic Modelling of Deterministic and Stochastic Oscillators*, Springer Series in Synergetics, https://doi.org/10.1007/978-3-031-81553-9_5

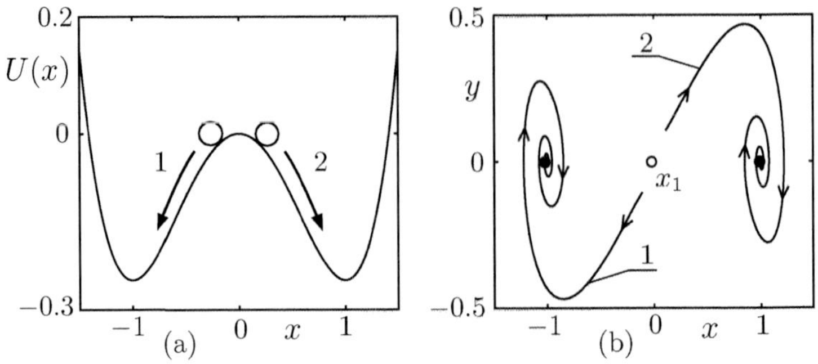

Fig. 5.1 Illustration of the potential function (panel (**a**)) and the phase portrait (panel (**b**)) of deterministic system (5.2) ($\xi(t) = 0$) in the regime of bistability. Parameters are: $m = 1$, $\gamma = 0.5$

$-m\frac{x^2}{2}+\frac{x^4}{4}$. In such a case, model (5.1) takes the form $\frac{d^2x}{dt^2} = -\gamma\frac{dx}{dt}+mx-x^3+\xi(t)$ and describes stochastic motion of a particle in a double-well potential field in the presence of dissipation (schematically shown in Fig. 5.1a). Introducing a new dynamical variable $y = \frac{dx}{dt}$, system (5.1) is rewritten as a system of two first-order ordinary differential equations:

$$\begin{cases} \frac{dx}{dt} = y, \\ \frac{dy}{dt} = -\gamma y + mx - x^3 + \xi(t). \end{cases} \tag{5.2}$$

For finite dissipation $\gamma > 0$ and in the absence of noise ($\xi(t) = 0$) the pitchfork bifurcation occurs in the system when increasing the parameter m. When m is negative, a single attractor, a stable steady state $x_1 = y_1 = 0$ exists in the phase space. The pitchfork bifurcation occurs at $m_* = 0$. After the bifurcation, the initially stable steady state becomes a saddle equilibrium and two stable steady states $x_{2,3} = \pm\sqrt{m}$, $y_{2,3} = 0$ appear in its vicinity. The eigenvalues characterising the appeared coexisting steady states are $\lambda_{2,3} = (-\gamma \pm \sqrt{\gamma^2 - 8m})/2$. The described transition from monostability to bistability is fully illustrated in the phase-parametric diagram depicted in Fig. 1.15 for the pitchfork bifurcation normal form. The only difference is the fact that model (5.2) is two-dimensional and its phase space is the phase plane (x, y). The trajectories tracing motion to coexisting stable steady states in the regime of bistability are illustrated in Fig. 5.1b for $m > 0$ and $0 < \gamma < 2\sqrt{2m}$, where the coexisting steady states are stable focuses. In such a case the phase trajectories trace spirals in their vicinities. In case $\gamma > 2\sqrt{2m}$ the attractors in the regime of bistability are stable nodes and the phase point directly moves to the attractors without tracing spiral-like trajectories.

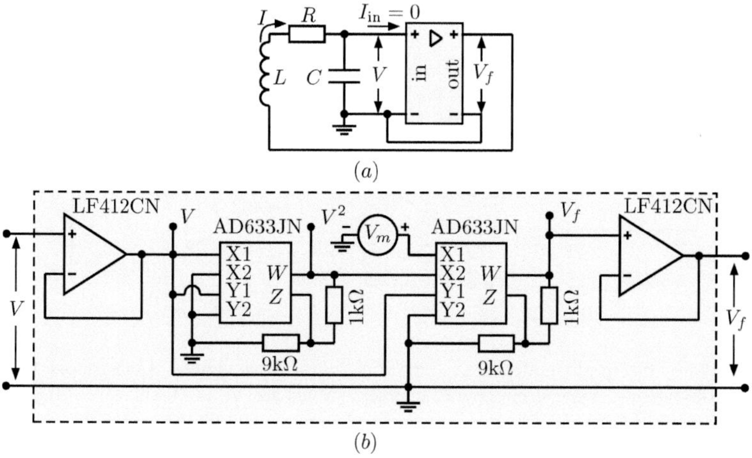

Fig. 5.2 Circuit implementation of model (5.2) by using the LCR-circuit with feedback: full circuit (panel (**a**)) and the nonlinear amplifier (panel (**b**))

To implement dynamical system (5.2) as an electronic circuit, one can apply two approaches. The first one is using the LCR-circuit with feedback as described in Chap. 2 (see Sect. 2.2.3 and the model development for the circuit in Fig. 2.16b. The corresponding circuit is depicted in Fig. 5.2a) where the series oscillatory circuit is forced by the nonlinear amplifier output voltage. The nonlinear amplifier circuit (illustrated in Fig. 5.2b) contains an operational-amplifier-based voltage follower and two analog multipliers. The nonlinear amplifier operates such that the output signal V_f is a cubic function of the input voltage (the voltage V across the capacitor C) $V_f = V(V_m - V^2)$. The used analog multipliers AD633JN have high (10 MΩ) input resistances which makes loading the LCR-circuit negligible. For this reason, it is not necessary to include the voltage follower at the nonlinear amplifier input. However, one can add the voltage follower based on an operational amplifier with high input impedance (for instance, the input impedance of the operational amplifier LF412CN is $10^{12}\,\Omega$), if it is principally important to maximally increase the nonlinear amplifier input resistance and, as a result, to minimize the input current and the impact on the LCR-circuit.

Using Kirchhoff's second law for the RLC-loop, $L\frac{dI}{dt} + IR + V = V_f(V)$, one can derive a second-order differential equation for the voltage V (see details in Sect. 2.2.3): $\frac{d^2V}{dt^2} + RC\omega_0^2\frac{dV}{dt} + \omega_0^2 V = \omega_0^2 V_f(V)$, where $\omega_0 = 1/\sqrt{LC}$. Introducing new time variable, $\tau = \omega t$, and the function V_f, the equation takes the form: $\frac{d^2V}{d\tau^2} = -R\sqrt{\frac{C}{L}}\frac{dV}{d\tau} + (V_m - 1)V - V^3$. Then the equation is rewritten for a dimensionless variable $x = V/1\text{Volt}$ and parameters $\gamma = -R\sqrt{\frac{C}{L}}$ and

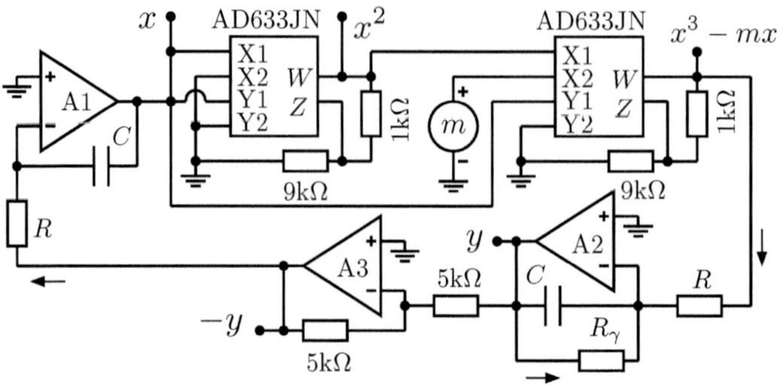

Fig. 5.3 Analog electronic model for system (5.2). All the input signals of integrators A1 and A2 are marked by arrows. Typical integrators' element values are: $R = 10\,\text{k}\Omega$, $C = 10\,\text{nF}$. Analog integrated circuits A1,A2 and A3 are the operational amplifiers TL072CP

$m = (V_m - 1)/1\text{Volt}$: $\frac{d^2x}{d\tau^2} = -\gamma\frac{dx}{d\tau} + mx - x^3$. After transition to a system of two first-order ordinary differential equation by introducing $\frac{dx}{dt} = y$ one obtains a deterministic modification of system (5.2) (without $\xi(t) = 0$).

The second approach for the implementation of system (5.2) is analog modelling. The corresponding circuit is depicted in Fig. 5.3. Here, signals x and y are the output voltages of integrators A1 and A2. The input signals for the integrator A1 is $-y$, while the inputs for the integrator A2 are y and $x^3 - mx$. Using the methodology described in Sect. 3.3, one can derive the differential equations for the circuit in Fig. 5.3.

$$\begin{cases} RC\dfrac{dx}{dt} = y, \\ RC\dfrac{dy}{dt} = -\gamma y + mx - x^3, \end{cases} \tag{5.3}$$

where $\gamma = \frac{R}{R_\gamma}$. All the terms in Eqs. (5.3) are voltages and the transition to the dimensionless form is carried out as division of all the equation terms by 1 Volt and introduction of dimensionless time $\tau = \frac{1}{RC}t$. Then the expressions for new derivatives are derived: $\frac{dx}{d\tau} = \frac{dx}{dt}\frac{dt}{d\tau} = \frac{dx}{dt}\frac{1}{RC}$ and $\frac{dy}{d\tau} = \frac{dy}{dt}\frac{1}{RC}$. It allows one to obtain the dimensionless form which fully coincides with system (5.2) for $\xi(t) = 0$.

5.1.2 *Overdamped Kramers Oscillator and Stochastic Resonance*

Introducing new time, $t_* = t/\gamma$, one transforms model (5.1) into the form $\frac{d^2x}{dt_*^2} + \frac{dx}{dt_*} = -\frac{dU(x)}{dx} + \xi(t_*)$. In the limit case $\gamma \to \infty$, the term $\frac{d^2x}{dt_*^2}$ tends to zero and the model becomes one-dimensional: $\frac{dx}{dt_*} = -\frac{dU(x)}{dx} + \xi(t_*)$. This model is obtained for extremely high dissipation and that is why it is called 'overdamped oscillator'. In the following, the overdamped oscillator is considered for a double-well potential function, $U(x) = -m\frac{x^2}{2} + \frac{x^4}{4}$, in the presence of stochastic and deterministic periodic force. To simplify the equation, the new time variable t_* is mentioned without the star. Then the overdamped Kramers oscillator takes the form:

$$\frac{dx}{dt} = mx - x^3 + \xi(t) + A\sin(\omega_{\mathrm{ext}}t), \tag{5.4}$$

where A and ω_{ext} are the amplitude and frequency of external periodic forcing. An analog model for system (5.4) is depicted in Fig. 5.4. Here, operational amplifier A2 is used to invert the external deterministic and stochastic input signals before integrator A1 processes them. The circuit can be simplified, if the mean values of the noise signal $\xi(t)$ and the periodic signal $A\sin(\omega_{\mathrm{ext}}t)$ are zero and the noise probability density function is assumed to be symmetric. In such a case, it is not necessary to use inverting amplifier A2 and one can include external signals by connection to integrator A1 through resistors R. The equation for the circuit in Fig. 5.4 is $RC\frac{dx}{dt} = mx - x^3 + \xi(t) + A\sin(\omega_{\mathrm{ext}}t)$.

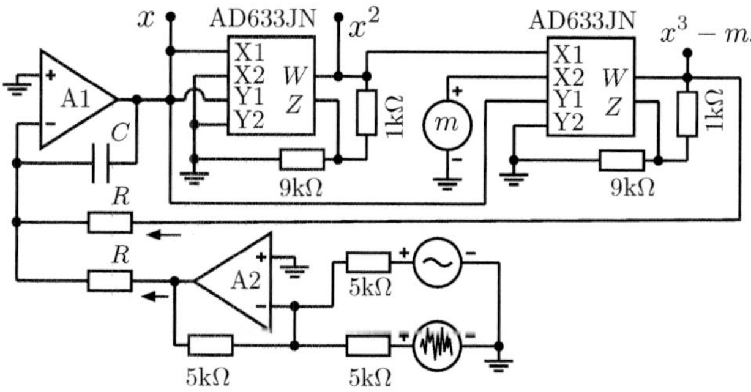

Fig. 5.4 Analog model for system (5.4). All the input signals of integrator A1 are marked by arrows. Typical integrator's element values are: $R = 10\,\mathrm{k\Omega}$, $C = 10\,\mathrm{nF}$. Analog integrated circuits A1 and A2 are the operational amplifiers TL072CP

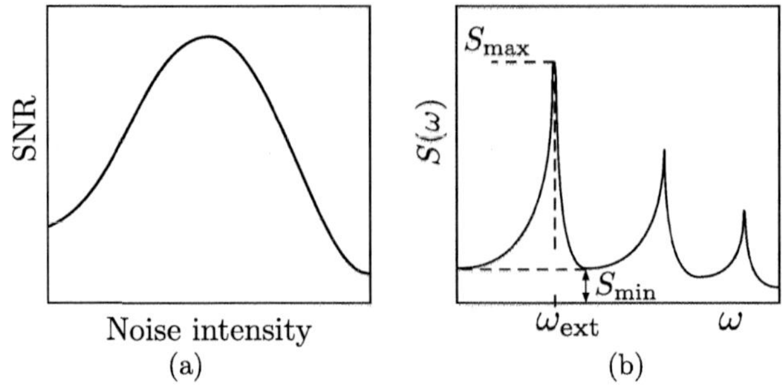

Fig. 5.5 (**a**) Dependence of the signal-to-noise ratio (SNR) on the noise intensity being typical for stochastic resonance; (**b**) Illustration for the SNR calculations

Model (5.4) became a paradigmatic model for the observation of the effect of stochastic resonance. Stochastic resonance is common in non-autonomous bistable dynamical systems and represents an interdisciplinary noise-induced phenomenon. This effect is manifested by increasing regularity of the stochastic dynamical system response to an input signal for increasing noise intensity. The effect of stochastic resonance is observed in a broad variety of non-autonomous dynamical systems and unites a broad spectrum of stochastic processes observed in the frameworks of laser physics [22–24], electronics [25–27], chemistry [28–33], biology [34–36], geophysics [37], climatology [19, 20, 38], population dynamics [39–41], economics [42]. This effect is widely used in practice (for instance, see Refs. [43–46] and the list of references in the reviews by L. Gammaitoni [47] and V. Anishchenko [48]). The issues addressing stochastic resonance are not limited by the single-system stochastic dynamics, but also involve the collective oscillatory behaviour [49–51], effects in spatially-extended systems [52–56] and time-delay oscillators considered as a spatially-extended system by means of virtual space-time representation [57]. This phenomenon can be observed in itself as well as be accompanied by effects of pattern formation such as spiral wave excitation [53] and chimera states [57].

The exhibition of stochastic resonance by bistable model (5.4) with varying intensity of the noise signal $\xi(t)$ can be illustrated by the non-monotonic dependence of the signal-to-noise ratio (SNR) at the external periodic signal frequency ω_{ext} on the noise intensity. The typical dependence of the SNR corresponding to the stochastic resonance is depicted in Fig. 5.5a. Here, one can notice increasing regularity of system's (5.4) response to an input signal with increasing noise intensity and the presence of local maximum of the function. This means that there exists the appropriate noise intensity level corresponding to the most regular system's response.

One can introduce the SNR in different forms. One of the possible options involves the evolution of power spectrum of oscillations $x(t)$ and is illustrated in Fig. 5.5b. The oscillations of a system driven by periodic forcing are characterised by the spectral peak S_{max} at the periodic signal frequency, ω_{ext}. In most cases, the spectral peak $S_{max} = S(\omega_{ext})$ is the main peak in the power spectrum. The power spectrum also has a minimum $S_{\min}$ close to the spectral peak. In radiophysics the most common definition of the signal-to-noise ratio is SNR $= P_S/P_N$, where P_S is the power of the signal and P_N is the noise power. The following formula of SNR corresponds to the harmonic external input signal in experiments: SNR $= H_s/H_n$, where H_s is the height of the spectral line above the background noise level in the power spectrum, and H_n is the background noise level close to the resonance frequency ω_{ext}, and thus in terms of the power spectrum SNR$=(S_{\max} - S_{\min})/S_{\min}$.

5.1.3 Double-Well Oscillator with Nonlinear Dissipation

Let us consider the bistable oscillator proposed in Refs. [58, 59] (schematically depicted in Fig. 5.6). Here, the circuit diagram contains two nonlinear elements N1 and N2 with the S- and N-type of the I-V characteristic, respectively: $I_{N1} = F(V)$, $V_{N2} = G(I)$. The circuit is similar to Nagumo's tunnel diode neuron model [60, 61], except of nonlinear resistor N2 connected in series with the inductor, L. The circuit also includes a source of broadband Gaussian noise current $I_{\text{noise}}(t)$, which will be assumed white in the following. By using the Kirchhoff's current law the following differential equations for the voltage V across the capacitor C and the current I through the inductor L can be derived:

$$\begin{cases} C\dfrac{dV}{dt'} + i_{N1} + I + I_{\text{noise}}(t') = 0, \\ V = L\dfrac{dI}{dt'} + V_{N2}. \end{cases} \tag{5.5}$$

In the dimensionless variables $x = V/v_0$ and $y = i/i_0$ with $v_0 = 1$ Volt, $i_0 =$ 1 Ampere and dimensionless time $t = [(v_0/(i_0 L)]t'$, Eq.(5.5) can be re-written as,

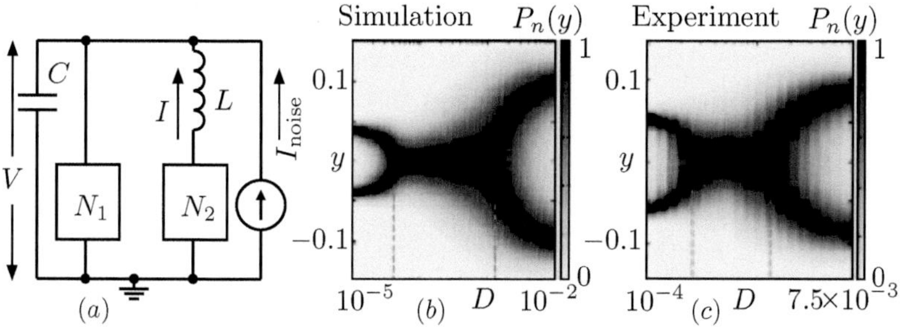

Fig. 5.6 (**a**) Schematic circuit diagram of model (5.5); (**b**)–(**c**) Evolution of the normalized probability density function caused by noise intensity, $P_n(y, D)$, obtained by numerical simulations of system (5.8) (panel (**b**)) and experimentally registered transformations (panel (**c**)) observed in analog model (5.12) (see Ref. [58] for details)

$$\begin{cases} \varepsilon \dfrac{dx}{dt} = -y - F(x) - \sqrt{2D}n(t), \\ \dfrac{dy}{dt} = x - G(y), \end{cases} \tag{5.6}$$

The parameter ε sets separation of slow and fast variables of the system, $\varepsilon = (C/L)(v_0/i_0)^2$. The first equation for the voltage contains an additive source of white Gaussian noise, $n(t)$, with the intensity D: $< n(t) = 0 >$, $< n(t)n(t+\tau) >= \delta(\tau)$. Depending on the shape of the functions $F(x)$ and $G(y)$ the circuit demonstrates wide range of dynamics, including various types of bistability, self-sustained oscillations and excitability. It allows to observe a wide range of dynamical regimes: from the behavior like in the Kramers oscillator with a double-well potential to dynamics of an excitable oscillator or a bistable self-sustained oscillators [62]. In addition, the development of model (5.6) made it possible to introduce a concept of deterministic [63] and noise-induced [64] self-oscillatory motions in a state-dependent potential well under the condition of positive dissipation. Moreover, studying asymmetric form of system (5.6) made it possible to develop a bistable ensemble model for the observation and control of spiking activity bump expansion [65]. This section is restricted to bistability of two stable equilibria with N-shape function $G(y) = -ay + by^3$ with positive coefficients a and b. The first circuit modification contains a linear resistor N1, $F(x) = c_1 x$, with positive c_1. Then circuit model (5.6) can be transformed into the following second-order differential equation:

$$\varepsilon \frac{d^2y}{dt^2} = -((3by^2 - a)\varepsilon + c_1)\frac{dy}{dt} - (1 - c_1 a + c_1 b y^2)y - \sqrt{2D}n(t). \tag{5.7}$$

The friction (dissipation) is nonlinear, but depends solely on the "coordinate" variable, y. For sufficiently small ε and $c_1 \gg \varepsilon(3by^2 - a)$, the dissipation becomes essentially linear. Then the system is closely akin to the Kramers oscillator with linear friction. As reported in Ref. [58], there is no qualitative differences in dynamics of the Kramers oscillator and model (5.7) with linear resistor N1.

Next, it is assumed that resistor N1 is nonlinear with $F(x) = c_1 x - c_3 x^3 + c_5 x^5$ and with fixed positive coefficients $c_{1,3,5}$. A variety of the electronic elements and circuits have I–V characteristic like that. For example, N1 can be realized by the so-called lambda-diode circuit [66]. In this case the following equations describes the system under study:

$$\begin{cases} \varepsilon \dfrac{dx}{dt} = -y - c_1 x + c_3 x^3 - c_5 x^5 - \sqrt{2D}n(t), \\ \dfrac{dy}{dt} = x + ay - by^3. \end{cases} \tag{5.8}$$

Equation (5.8) can be written in the "coordinate-velocity" form with dynamical variables y, $v \equiv \dfrac{dy}{dt}$ and $x = v - ay + by^3$:

$$\begin{cases} \dfrac{dy}{dt} = v, \\ \varepsilon\dfrac{dv}{dt} = -y - c_1(v - ay + by^3) + c_3(v - ay + by^3)^3 - \\ c_5(v - ay + by^3)^5 + \varepsilon v(a - 3by^2) - \sqrt{2D}n(t). \end{cases} \tag{5.9}$$

In the oscillatory form (5.9) becomes,

$$\ddot{y} + q_1(y, \dot{y})\dot{y} + \frac{1}{\varepsilon}q_2(y) = -\sqrt{2D}n(t), \tag{5.10}$$

where $\ddot{y} = \dfrac{d^2y}{dt^2}$, $\dot{y} = \dfrac{dy}{dt}$, $q_2(y)$ defines the form of the potential, and $q_1(y, \dot{y})$ is the nonlinear dissipation,

$$\begin{aligned} q_1(y, \dot{y}) &= -a + 3by^2 + \frac{1}{\varepsilon}(c_1 - \\ & c_3 \sum_{n=1}^{3} \tfrac{3!}{n!(3-n)!} \dot{y}^{n-1}(by^2 - a)^{3-n} y^{3-n} + \\ & c_5 \sum_{n=1}^{5} \tfrac{5!}{n!(5-n)!} \dot{y}^{n-1}(by^2 - a)^{5-n} y^{5-n}), \\ q_2(y) &= y + c_1(by^2 - a)y - c_3(by^2 - a)^3 y^3 + c_5(by^2 - a)^5 y^5. \end{aligned} \tag{5.11}$$

It must be noted that unlike for the case of linear resistor Eq.(5.7), the system's dissipation now depends on both coordinate and velocity, y and $\dot{y}$. As shown in Ref. [58], it gives the main reason of the qualitative difference between dynamics of system (5.9) and the behaviour of models (5.1) and (5.7).

One of the interesting phenomena exhibited by model (5.9) are noise-induced transitions 'bistability-monostability-bistablility' caused by increasing noise intensity [58]. These transitions are illustrated as the transformation of the probability density function when varying noise level: from bimodal distribution to the unimodal and then new transition back to the bimodal distribution. In particular, the discussed evolution of the normalized probability function $P_n(y) = P(y)/P_{\max}(y)$ obtained in numerical simulations of Eqs. (5.9) is depicted in Fig. 5.6b. Qualitatively same results were obtained in physical experiments with the corresponding analog model (see Fig. 5.6c).

The circuit of analog model developed for systems (5.8) and (5.9) is presented in Fig. 5.7. Here the operational amplifier-based integrators are A1 and A4, their output voltages are taken as the dynamical variables, x and y, respectively. Then the signals x and y are transformed in order to match expressions of the right hand side of Eqs. (5.8). The necessary signal transformations are carried out by using analog multipliers AD633JN and the operational amplifiers TL072CP connected in the inverting and non-inverting amplifier configurations. Finally, transformed signals come to the input of the integrators. The experimental setup allows to obtain the

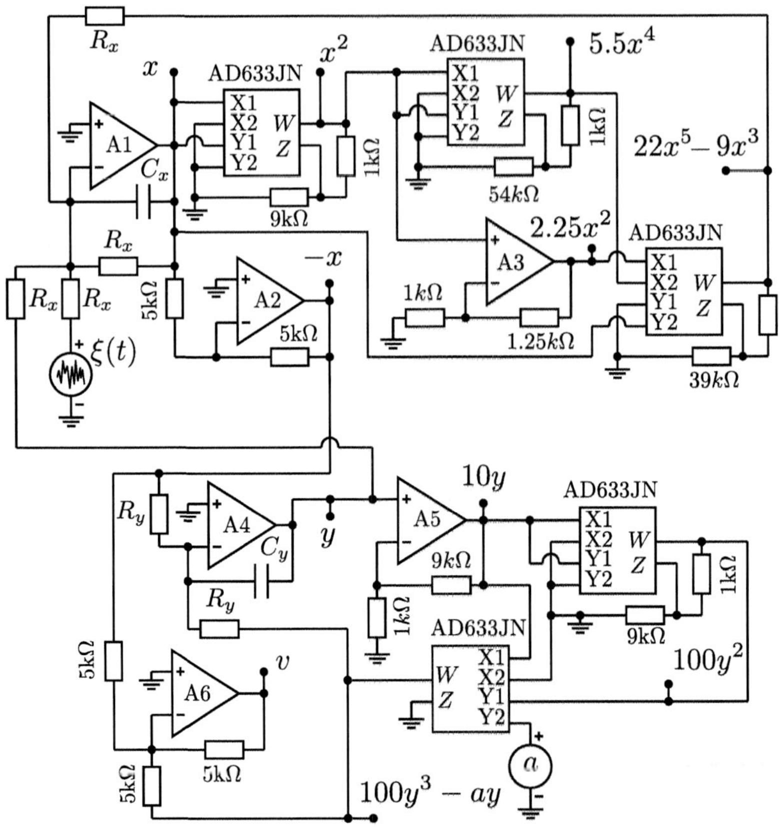

Fig. 5.7 Analog electronic model for system (5.8). Analog integrated circuits A1–A6 are the operational amplifiers TL072CP. Parameters are: $R_x = 1\,\text{k}\Omega$, $R_y = 10\,\text{k}\Omega$, $C_x = 30\,\text{nF}$, $C_y = 300\,\text{nF}$

instantaneous values of the variables x, y and $v = \dot{y} = x + ay - by^3$. The circuit in Fig. 5.7 is described by the following equations:

$$\begin{cases} R_x C_x \dfrac{dx}{dt} = -y - x + 9x^3 - 22x^5 - \xi(t), \\ R_y C_y \dfrac{dy}{dt} = x + ay - 100y^3, \end{cases} \tag{5.12}$$

where $C_x = 30$ nF, $C_y = 300$ nF, $R_x = 1$ kΩ is the resistance at the integrator A1, $R_y = 10$ KΩ is the resistance at the integrator A10. The parameter a is equal to the input value of the voltage $a = 1.2$ V at the analog multiplier. Transition to dimensionless equations (5.8) is then carried out by substitution dimensionless time $\tau = t/R_0C_0$ and variables $x_* = x/V_0$, $y_* = y/V_0$, $v_* = v/V_0$, where $R_0C_0 =$

$R_y C_y = 3$ ms is the circuit's time constant and $V_0 = 1$ Volt. Then the dimensionless parameter ε takes the value: $\varepsilon = \dfrac{R_x C_x}{R_y C_y} = 0.01$.

5.2 Self-Oscillations

Generally speaking, self-oscillation (are often called self-sustained oscillations) is a kind of periodic motion by a source of power that lacks any corresponding periodicity. The oscillator itself controls the phase, frequency and amplitude. Self-oscillators therefore differs from forced and parametric resonators, where the power that sustains the motion must be externally modulated. The problem of the self-sustained oscillation existence in autonomous dynamical systems was originally introduced by H. Poincare [67]. Thereafter bifurcation mechanisms of self-sustained oscillation excitation were described by A.Andronov and his associates [68] and E. Hopf [69, 70]. Despite the issue of the self-sustained oscillations is known for over a century, this topic is still attractive and interesting. This is due to the fact that self-sustained oscillators are of a frequent occurrence in physics [71–74], chemistry [75–77], geology [78, 79], climatology [80–82], biology [83–86], economics [87, 88] and other fields. Problematics of self-oscillations in deterministic systems includes regular periodic motions as well as the chaotic dynamics [89–91]. There are well-known stochastic effects related to the self-oscillation topic: stochastic bifurcations (in the context of the stochastic Andronov-Hopf bifurcation [92–97]), coherence resonance [98–103], stochastic synchronization [104–106].

In the further sections, the issue of self-oscillators is explored on examples of Van der Pol self-sustained oscillator and the Brusselator model. In addition, the self-oscillations are discussed in Sect. 5.3 in the context of the FithHugh-Nagumo model and in Sect. 5.4 on examples of the Anishchenko-Astakhov self-oscillator and the Rössler oscillator.

5.2.1 *Van der Pol Self-Oscillator*

The van der Pol self-sustained oscillator was proposed in 1920 to explain stable oscillations (called by Balthasar van der Pol as relaxation-oscillations) in the triode-based circuit [107]. The autonomous Van der Pol system exhibits the regular quiescent or self-oscillatory dynamics, while the non-autonomous model can realize the regime of deterministic chaos: van der Pol and van der Mark reported that an irregular noise was heard at certain frequencies of external periodic forcing [108] which was later found to be the result of deterministic chaos. In the simplest form, the van der Pol model is described by the following equation:

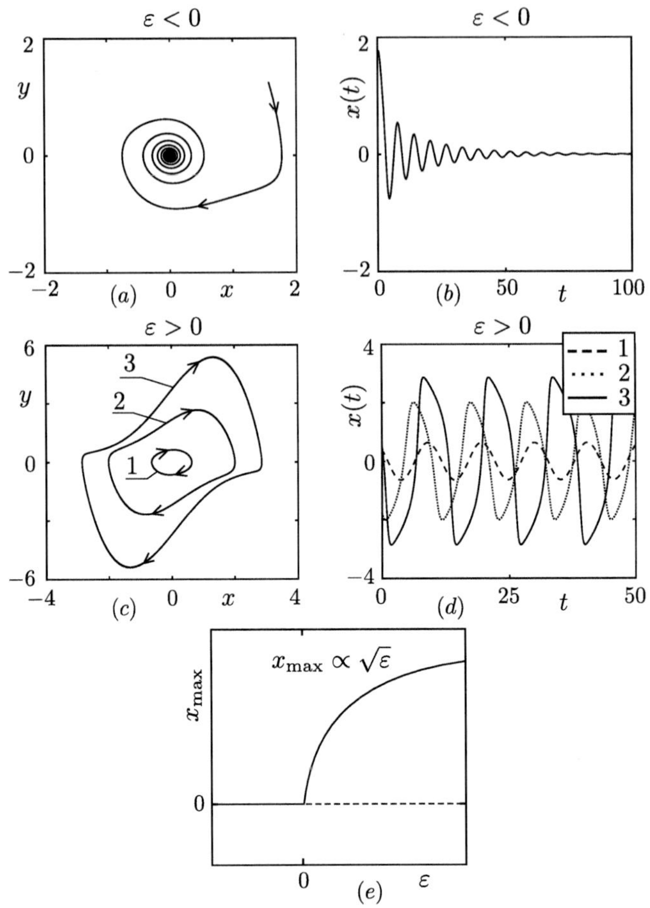

Fig. 5.8 Phase portraits on the plane (x, $y = \frac{dx}{dt}$) and time realizations $x(t)$ for system (5.13) at $\varepsilon = -0.1$ (panels (**a**) and (**b**)) and at positive ε (panels (**c**) and (**d**)): $\varepsilon = 0.1$ (closed curve 1), $\varepsilon = 0.5$ (closed curve 2), $\varepsilon = 1.0$ (closed curve 3). Panel (**e**) schematically illustrates the dependence of the oscillation amplitude x_{max} on parameter ε being typical for systems exhibiting soft self-oscillation excitation through the supercritical Andronov-Hopf scenario

$$\frac{d^2x}{dt^2} - (\varepsilon - x^2)\frac{dx}{dt} + x = 0, \tag{5.13}$$

where x is a dynamical variable, ε is a parameter, which determines the dynamics. The phase portraits of the system for positive and negative values ε are depicted in Fig. 5.8a, c. The phase space evolution at $\varepsilon = 0$ results from the bifurcation transition corresponding to the soft self-oscillation excitation. Before the bifurcation ($\varepsilon < 0$) the system's attractor is a stable steady state in the origin (Fig. 5.8a) and realizations $x(t)$ trace damped oscillations (Fig. 5.8b). After the supercritical

Andronov-Hopf bifurcation occurring at $\varepsilon = 0$ the fixed point becomes unstable and a stable limit cycle appears in its vicinity. For sufficiently small positive values of ε the impact of nonlinearity is weak and the self-oscillations are well-described as $x(t) = x_{\max} \cos(t)$ (see the solution 1 in Fig. 5.8c, d for $\varepsilon = 0.01$). For large ε the influence of system's nonlinearity is strong and one observes nonlinear oscillations such solutions 2 and 3 in Fig. 5.8c, d for $\varepsilon = 0.5$ and $\varepsilon = 1$.

The bifurcation transition in model (5.13) observed at $\varepsilon = 0$ demonstrates fundamental properties of soft oscillation excitation. For positive ε, but close to the supercritical Andronov-Hopf bifurcation, the self-oscillation amplitude $x_{\max}$ continuously depends on the parameter ε and is proportional to $\sqrt{\varepsilon}$ (see Fig. 5.8e). This results in the fact that one can achieve any amplitudes in system (5.13) including extremely low. The dependence of the amplitude as a function of a square root of the controlling parameter represents a universal peculiarity exhibited by self-oscillators where the supercritial Andronov-Hopf bifurcation is realized.

Consider the van der Pol self-oscillator in the presence of fifth-order nonlinearity:

$$\frac{d^2x}{dt^2} - (\varepsilon + \mu x^2 - x^4)\frac{dx}{dt} + x = 0, \tag{5.14}$$

where one deals with the second parameter, $\mu > 0$. Increasing parameter ε gives rise to hard oscillation excitation. First, growth of the parameter ε results in the occurrence of the saddle-node bifurcation of limit cycle. Before the bifurcation, a single attractor, a stable steady state exists in the phase space (Fig. 5.9a). After the saddle-node bifurcation occurring at $\varepsilon = -\dfrac{\mu^2}{8}$ stable and unstable limit cycles appear in the phase space and one observes the regime of bistability: there are two attractors, a stable fixed point and a stable limit cycle. The basins of attraction are separated by an unstable limit cycle repelling the phase trajectories (Fig. 5.9b). Further increasing parameter ε leads to the contraction of the unstable limit cycle which collapses in the steady state in the origin at $\varepsilon = 0$. This phase space transformation corresponds to the subcritical Andronov-Hopf bifurcation. In case $\varepsilon > 0$ one observes the regime of monostability again: after the subcritical Andronov-Hopf bifurcation the steady state becomes unstable and the stable limit cycle is a single attractor in the phase space (Fig. 5.9c).

The phase-parametric diagram in Fig. 5.9d obtained for model (5.14) describes the hard self-oscillation excitation: the oscillation amplitude rapidly increases after passing the critical value $\varepsilon = 0$ and it is impossible to achieve extremely low amplitudes. However, if one keeps the phase point to move along the limit cycle and slowly decrease the parameter ε, then damping self-oscillation occurs at $\varepsilon = -\dfrac{\mu^2}{8}$. The described phenomenon is called hysteresis: the self-oscillation excitation and damping occur at different parameter values.

Equations (5.13) and (5.14) became classical models for the demonstration of self-oscillations and describes a wide range of oscillatory processes. In particular, they describe a broad variety of electronic circuits. Three of them are depicted in Fig. 5.10. The equations for the circuit in Fig. 5.10a are (for details, see the

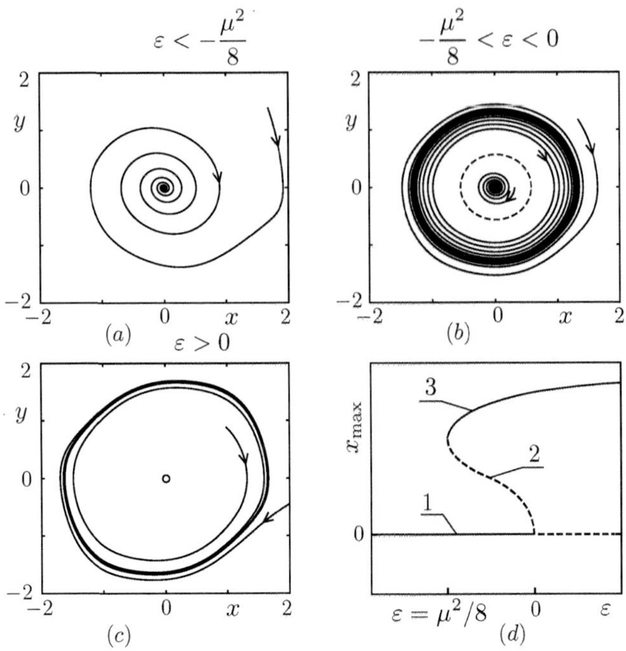

Fig. 5.9 Phase portraits on the plane $(x,\ y = \frac{dx}{dt})$ of system (5.14) at $\varepsilon = -0.1$ (panel (**a**)), $\varepsilon = -0.01$ (panel (**b**)), $\varepsilon = 0.1$ (panel (**c**)). Panel (**d**) schematically illustrates the dependence of the oscillation amplitude $x_{\max}$ on parameter ε. Curve 1 corresponds to the equilibrium point (stable for $\varepsilon < 0$ and unstable for $\varepsilon > 0$), while curves 2 and 3 mark unstable and stable limit cycles

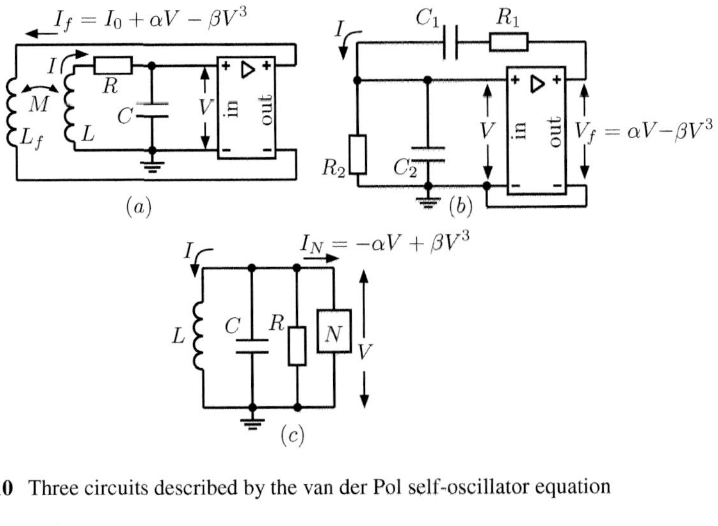

Fig. 5.10 Three circuits described by the van der Pol self-oscillator equation

description of the circuit in Fig. 2.16c presented in Sect. 2.2.3):

$$\frac{d^2V}{dt^2} + \left(\frac{R}{L} - \frac{M}{LC}\frac{dI_f(V)}{dV}\right)\frac{dV}{dt} + \omega_0^2 V = 0, \tag{5.15}$$

where $\omega_0 = \frac{1}{\sqrt{LC}}$. Substituting $I_f = I_0 + \alpha V - \beta V^3$, and introducing new time scale $\tau = \omega_0 t$, one obtains the system equation:

$$\frac{d^2V}{d\tau^2} - \left(\frac{\alpha M - RL}{\sqrt{LC}} - \frac{3M\beta}{\sqrt{LC}}V^2\right)\frac{dV}{d\tau} + V = 0, \tag{5.16}$$

which is finally transformed into form (5.13) by substitution of a dimensionless dynamical variable $x = \sqrt{\frac{3M\beta}{\sqrt{LC}}}V$ and introduction of a system parameter $\varepsilon = \frac{\alpha M - RL}{\sqrt{LC}}$.

A general model for the Wien-bridge-based circuit depicted in Fig. 5.10b is developed in Sect. 2.3 for any function V_f. In particular case $V_f = \alpha V - \beta V^3$ and $C_1 = C_2 = C$, $R_1 = R_2 = R$ the model equation takes the form

$$\frac{d^2V}{dt^2} + \omega_0\left(3 - \alpha + 3\beta V^2\right)\frac{dV}{dt} + \omega_0^2 V = 0, \tag{5.17}$$

where $\omega_0 = \frac{1}{CR}$. Similarly to the previous circuit in Fig. 5.10a, Eq. (5.17) is transformed into dimensionless form (5.13) by substitution of new time $\tau = \omega_0 t$ and introduction of a dynamical variable $x = \sqrt{3\beta}V$ and $\varepsilon = \alpha - 3$.

The equation for parallel oscillatory circuit in Fig. 5.10c for the cubic current voltage characteristic of element N, $I_N = -\alpha V + \beta V^3$, becomes (to derive the equation, see the description of circuit in Fig. 2.15b in Sect. 2.7):

$$\frac{d^2V}{dt^2} + \frac{1}{C}\left(\frac{1}{R} - \alpha + 3\beta V^2\right)\frac{dV}{dt} + \omega_0^2 V = 0, \tag{5.18}$$

where $\omega_0 = \frac{1}{\sqrt{LC}}$. Substituting the system parameter $\varepsilon = \left(\alpha - \frac{1}{R}\right)\sqrt{\frac{L}{C}}$ and dynamical variable $x = \sqrt{3\beta\sqrt{\frac{L}{C}}}V$ and rewriting the system for the dimensionless time $\tau = \omega_0 t$, one can transform Eq. (5.18) into form (5.13).

The models of the circuits in Fig. 5.10 are developed for the presence of cubic nonlinearity and are reduced to the same model, Eq. (5.13). Considering the circuits for the fifth-order polynomial nonlinearity, one can derive the equations and transform them into form (5.14).

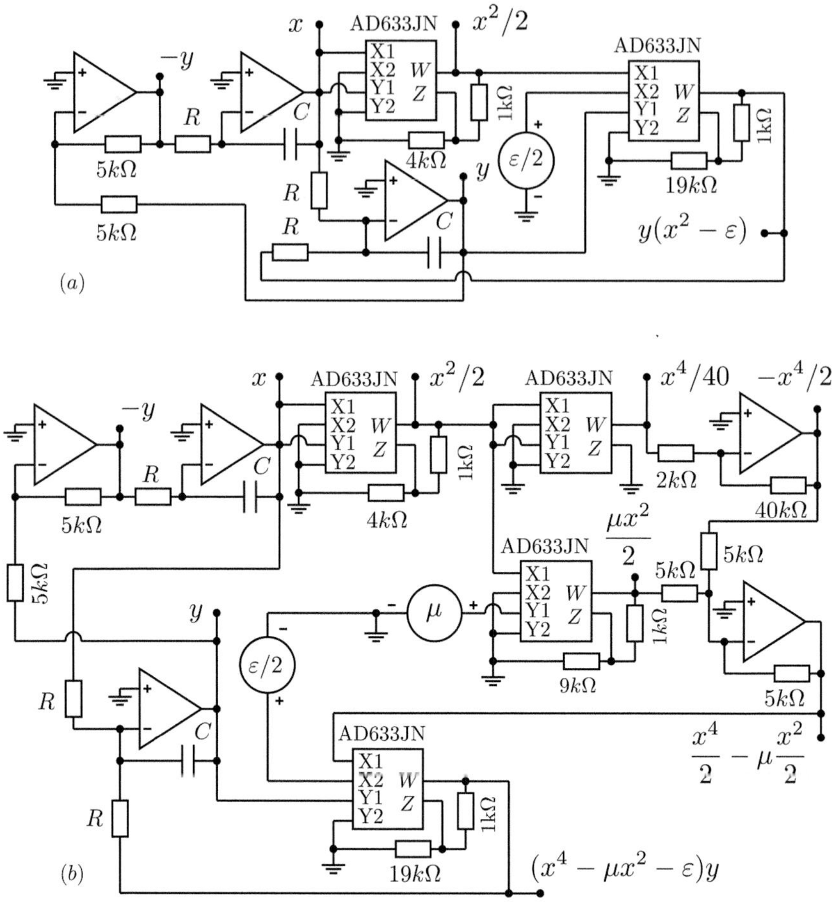

Fig. 5.11 Analog model developed for systems (5.13) (panel (**a**)) and (5.14) (panel (**b**)). In both circuits: $R = 10\,\text{k}\Omega$, $C = 10\,\text{nF}$, used operational amplifiers are TL072CP

Figure 5.11 shows the analog electronic models of dynamical systems (5.13) and (5.14) used for experimental study of stochastic Andronov-Hopf bifurcation [109–111]. The circuit equations are:

$$\begin{cases} RC\dfrac{dx}{dt} = y, \\ RC\dfrac{dy}{dt} = (\varepsilon - x^2)y - x, \end{cases} \tag{5.19}$$

and

$$\begin{cases} RC\dfrac{dx}{dt} = y, \\ RC\dfrac{dy}{dt} = (\varepsilon + \mu x^2 - x^4)y - x, \end{cases} \tag{5.20}$$

In both cases, the parameters ε and μ are determined by the analog multiplier input voltage. It provides for using a DC-voltage source to keep the parameter to be fixed, as well as to apply a periodic or stochastic signal sources. This feature made be possible to consider the impact of parametric noise on the Andronov-Hopf bifurcation besides the additive noise. This is one of the analog modelling advantages: all the terms in Eqs. (5.19) and (5.20) are voltages and it is much easier to apply the stochastic or deterministic external forcing to any block of the analog model circuit as compared to the circuits in Fig. 5.10.

5.2.2 *Brusselator*

The brusselator is a theoretical model for a type of autocatalytic reaction developed by the Nobel laureate Ilya Prigogine and associators in Brussels [112]. For many decades the brusselator model has been under study in the context of oscillatory dynamics and pattern formation far from thermodynamic equilibrium. In particular, the brusselator well describes a class of chemical reactions called the Belousov-Zhabotinsky reaction. In the single-oscillator form the brusselator's dynamics is determined by the following ordinary differential equations:

$$\begin{cases} \dfrac{dx}{dt} = A - x - Bx + x^2 y, \\ \dfrac{dy}{dt} = Bx - x^2 y. \end{cases} \tag{5.21}$$

There exists a fixed point with coordinates $(A, B/A)$ in the phase space of the brusselator. The steady state is stable at $B < 1 + A^2$ (Fig. 5.12a). Then increasing the parameter B causes the supercritical Andronov-Hopf bifurcation at $B = 1 + A^2$. After the bifurcation, the steady state becomes unstable and system's attractor is a stable limit cycle (Fig. 5.12b).

System (5.21) can be electronically implemented by means of analog modelling. The corresponding analog circuit is depicted in Fig. 5.13 and is described by the equations:

$$\begin{cases} RC\dfrac{dx}{dt} = A - x - Bx + x^2 y, \\ RC\dfrac{dy}{dt} = Bx - x^2 y. \end{cases} \tag{5.22}$$

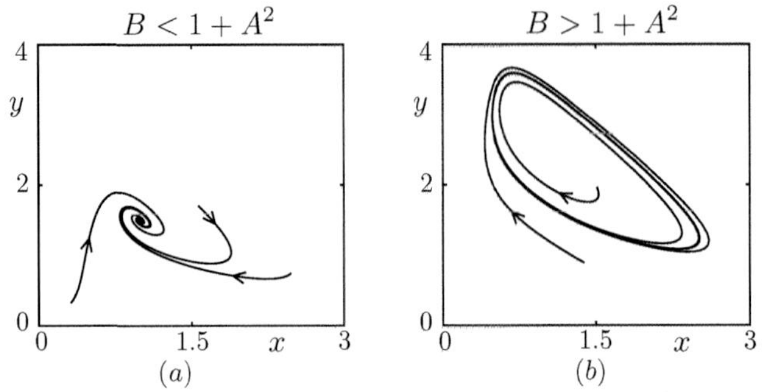

Fig. 5.12 Phase portrait of system (5.21) before (panel (**a**)) and after (panel (**b**)) the supercritical Andronov-Hopf bifurcation caused by increasing the parameter B

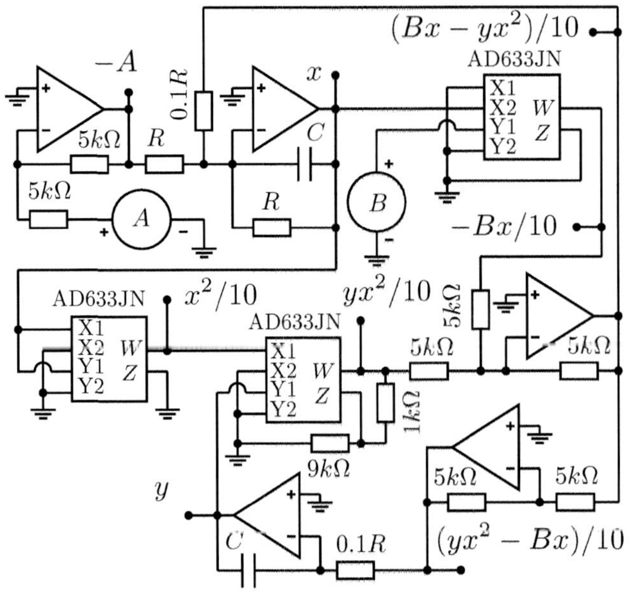

Fig. 5.13 Analog model developed for the brusselator model (5.21), where $R = 10\,\mathrm{k}\Omega$, $C = 10\,\mathrm{nF}$. Used operational amplifiers are TL072CP

Similarly to the van der Pol self-oscillator analog models in Fig. 5.11, the parameters A and B of the circuit in Fig. 5.13 are determined by the input voltages which can contain deterministic and stochastic signal components. This property was used to experimentally study the impact of parametric noise on the stochastic Andronov-Hopf bifurcation in the brusselator [113].

The second interesting detail is the fact that the nonlinear blocks of the circuit in Fig. 5.13 produce the signals $Bx/10$, $x^2y/10$ and $\pm\,(Bx - x^2y)/10$ which are the input signals for the integrators. Integrators' input resistances being responsible for these input signals are 0.1 R. For this reason, the terms $\pm\,(Bx - x^2y)$ persist in resulting Eq. (5.22) without division by 10. This feature allows to avoid extremely high amplitudes associated with signals Bx, x^2y and $\pm\,(Bx - x^2y)$.

5.3 Coherence Resonance and Excitability

The phenomenon of coherence resonance was originally discovered for excitable systems [98, 99, 114–117] and then for non-excitable ones [100–102, 106, 118]. This effect implies that noise-induced oscillations become more regular with increasing the noise intensity in a finite range of the noise level. As a result, there is an appropriate noise intensity corresponding to the most regular stochastic oscillations. Coherence resonance is a frequent occurrence in neurodynamics [98, 99, 119, 120] as well as in microwave [121] and semiconductor [122, 123] electronics, optics [124–128], thermoacoustics [129], plasma physics [130], and chemistry [131–133].

The noise-induced dynamics of excitable systems in the regime of coherence resonance is similar to the self-oscillatory behaviour. The similarity is complemented by the fact that noise-induced oscillations corresponding to the coherence resonance can be synchronized mutually or by external forcing [134–136]. Moreover, the synchronization of the noise-induced oscillations occurs in a similar way as for a deterministic quasiperiodic system [137].

In the further section, the effect of coherence resonance is demonstrated in details on an example of the FitzHugh-Nagumo oscillator.

5.3.1 Deterministic FitzHugh-Nagumo Oscillator

The FitzHugh-Nagumo oscillator represents is a simplified 2D version of the Hodgkin-Huxley model which describes the spiking neuron dynamics. Originally proposed for the description of processes in nerve membranes [60, 138], the FitzHugh-Nagumo system became a paradigmatic model for the type-II excitability and coherence resonance in excitable systems. In the current section, study of the FitzHugh-Nagumo model is started from the simplest deterministic form:

$$\begin{cases} \varepsilon \dfrac{dx}{dt} = x - x^3/3 - y, \\ \dfrac{dy}{dt} = x + a, \end{cases} \tag{5.23}$$

where $x =$ is the activator variable, y is the inhibitor variable. A parameter $\varepsilon \ll 1$ is responsible for the time scale separation of fast activator and slow inhibitor variable, a is the threshold parameter which determines the system dynamics: the system exhibits the excitable regime at $|a| > 1$ and the oscillatory one for $|a| < 1$.

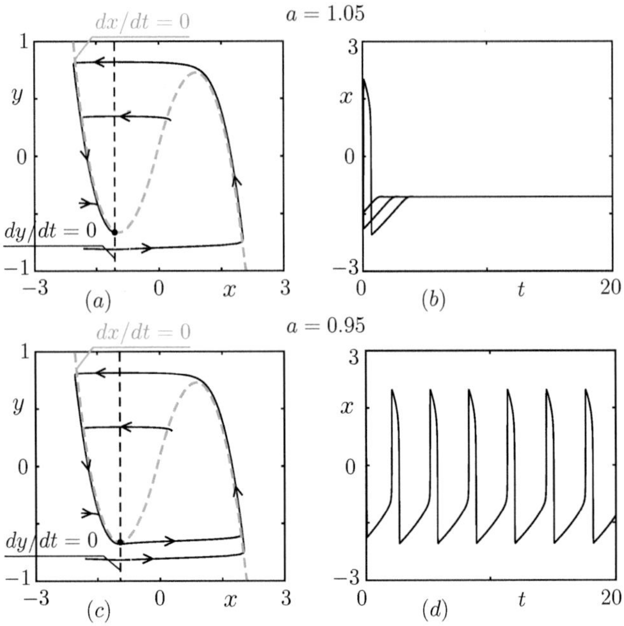

Fig. 5.14 Phase space and time realizations $x(t)$ of deterministic FitzHugh-Nagumo model (5.23) at $a = 1.05$ (panels (**a**) and (**b**)) and $a = 0.95$ (panels (**c**) and (**d**)). The second parameter is $\varepsilon = 0.01$. The nullclines $\frac{dx}{dt} = 0$ and $\frac{dy}{dt} = 0$ are marked by the dashed lines

In case $|a| > 1$ system (5.23) is characterised by the existence of a stable steady state in the phase space (Fig. 5.14a). The time realizations illustrate damped oscillations (Fig. 5.14b). Decreasing the absolute value $|a|$ induces the supercritical Andronov-Hopf bifurcation at $|a| = 1$. Then in case $|a| > 1$ system's attractor is a stable limit cycle corresponding to slow-fast self-oscillatory dynamics (Fig. 5.14b). It includes slow motions along the nullcline (the dashed curve in Fig. 5.14b corresponding to the condition $\frac{dx}{dt} = 0$) and the fast ones, when the phase point falls down from the nullcline. The corresponding time realizations $x(t)$ represent periodic impulse-like self-oscillations called spikes (Fig. 5.14d).

The considered modification of the FitzHugh-Nagumo oscillator exhibits soft self-oscillation excitation associated with the supercritical Andronov-Hopf bifurcation. However, there are more complicated forms of the FitzHugh-Nagumo model which can realize the occurrence of the hard self-oscillation excitation through the subcritical Andronov-Hopf bifurcation [139, 140].

5.3.2 Coherence Resonance in a Stochastic FitzHugh-Nagumo Model

Consider the FitzHugh-Nagumo oscillator in the presence of noise:

$$\begin{cases} \varepsilon \dfrac{dx}{dt} = x - x^3/3 - y, \\ \dfrac{dy}{dt} = x + a + \sqrt{2D}n(t), \end{cases} \tag{5.24}$$

where the second equation contains an additive source of white Gaussian noise, $n(t)$, with the intensity D: $< n(t) = 0 >$, $< n(t)n(t+\tau) >= \delta(\tau)$. Parameter a is fixed in the excitable regime, $a = 1.05$. The chosen parameter value corresponds to the existence of a stable fixed point in the phase space of the deterministic system (see Fig. 5.14a). In the presence of noise the random force occasionally kicks the phase point out of the vicinity of the stable equilibria. If a stochastic perturbation is sufficient, the phase point traces motion along a loop including a stable branch of the nullcline $\dfrac{dx}{dt} = 0$ (Fig. 5.15a). The motion along the loop is manifested as spikes in the $x(t)$ time realization (Fig. 5.15b). It must be noted that the term

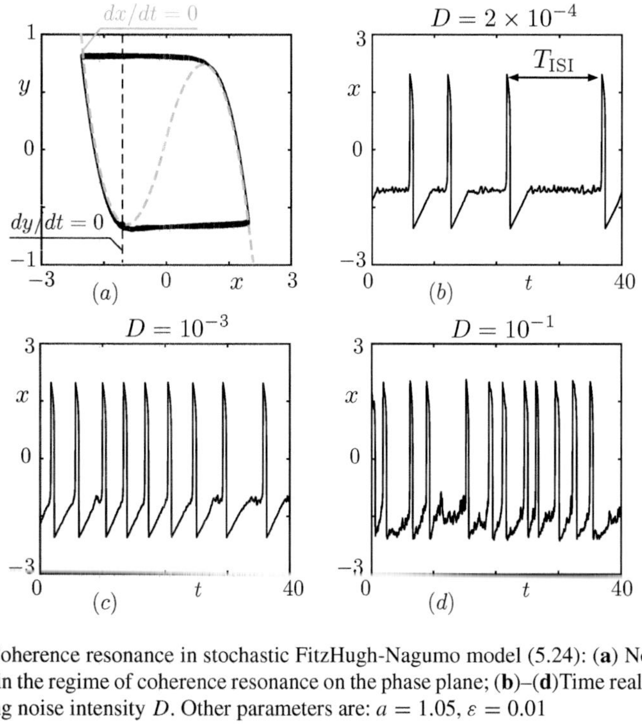

Fig. 5.15 Coherence resonance in stochastic FitzHugh-Nagumo model (5.24): (**a**) Noise-induced oscillations in the regime of coherence resonance on the phase plane; (**b**)–(**d**)Time realizations $x(t)$ for increasing noise intensity D. Other parameters are: $a = 1.05$, $\varepsilon = 0.01$

'excitability' is associated with the fact that the small-amplitude (which means small-energy) external perturbation can induce system's response with much higher energy (the system gets excited). Small noise (Fig. 5.15b) results in rare spikes with variable interspike intervals. Growth of the noise intensity gives rise to more frequent spiking and increasing the spiking regularity. There is an optimal value of the noise intensity corresponding to the most regular spiking (Fig. 5.15c) and the almost periodic dynamics. Further increasing the noise intensity leads to the destruction of the oscillation regularity (Fig. 5.15d).

In the regime of most regular noise-induced oscillations (see Fig. 5.15c) system's dynamics maximally resembles the deterministic self-oscillatory behaviour (compare the phase portraits in Figs. 5.14c and 5.15a and the time realizations in Figs. 5.14d and 5.15c). In terms of power spectra, coherence resonance is associated with the transformation of a spectrum such that the main spectral peak at the frequeency ω_0 being very close to the deterministic self-oscillation frequency becomes most prominent for an optimal noise intensity (see Fig. 5.16a) and then fades out with the noise-intensity growth. To characterise this quantitatively, one can introduce the correlation time t_{cor} as a quantity being proportional to S_{max}/w_p where S_{max} is the main peak hight and w_p is its width at the half-power level (schematically illustrated in Fig. 5.16a). The second approach involves the autocorrelation function for the correlation time characterisation. In the frameworks of this method, the correlation time is defined as

$$t_{cor} = \frac{1}{\sigma^2} \int_0^{\infty} |\Psi(s)|\, ds, \tag{5.25}$$

where $\Psi(s)$ is the autocorrelation function of the signal $x(t)$ and $\sigma^2 = \Psi(0)$ is its variance.

The second important characteristic for revealing coherence resonance is the normalized standard deviation of interspike intervals [98] (see the illustration of

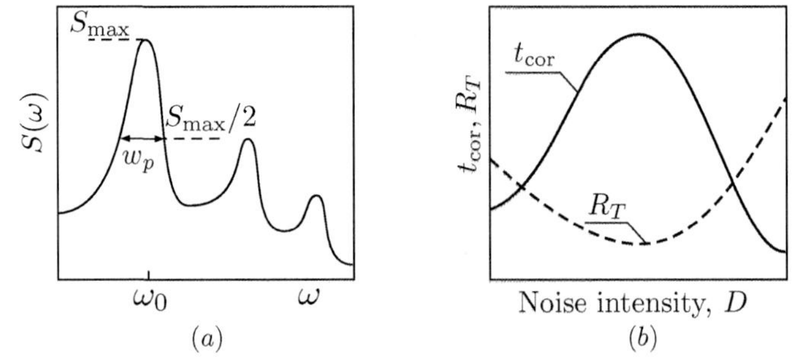

Fig. 5.16 (**a**) Schematic illustration of the noise-induced oscillation power spectra in the regime of coherence resonance; (**b**) Dependences of the correlation time t_{cor} and the deviation of interspike intervals R_T on the noise intensity D being typical for coherence resonance

the interspike interval in Fig. 5.15b). The standard deviation of interspike intervals is calculated as

$$R_T = \frac{\sqrt{\langle T_{ISI}^2 \rangle - \langle T_{ISI} \rangle^2}}{\langle T_{ISI} \rangle}. \tag{5.26}$$

The dependences $R_T(D)$ and $t_{\rm cor}(D)$ being typical for coherence resonance indicate the existence of an optimal noise intensity D_{opt} corresponding to a local maximum of the function $t_{\rm cor}(D)$ and a local minimum of the function $R_T(D)$ (see Fig. 5.16b).

5.3.3 FitzHgh-Nagumo Oscillator Circuit Implementations

The electronic circuit proposed by Jin-ichi Nagumo and associators for the implementation of the simplified Hodgkin-Huxley model represents a solution where the nonlinear properties are associated with the tunnel-diode current-voltage characteristic [60]. Technically, a circuit with similar properties (and described by similar equations) can be realized using different nonlinear one-port elements. For instance, such circuit with the operation-amplifier-based negative resistance is proposed in Ref. [111]. An oscillatory circuit for the FitzHugh-Nagumo implementation is generally illustrated in Fig. 5.17. Using Kirchhoff's laws for the circuit in Fig. 5.17, one can derive the system of two first-order differential equations for the voltage across the capacitor, V, and the current passing through the inductor and resistor, I: $I = C\dfrac{dV}{dt} + I_N(V)$ and $L\dfrac{dI}{dt} = \varepsilon - RI - V$. Next, it is assumed that the current through the nonlinear block depends as a cubic function on the voltage across the block, $I = -\alpha V + \beta V^3$. Then the model equations in physical variables take the form:

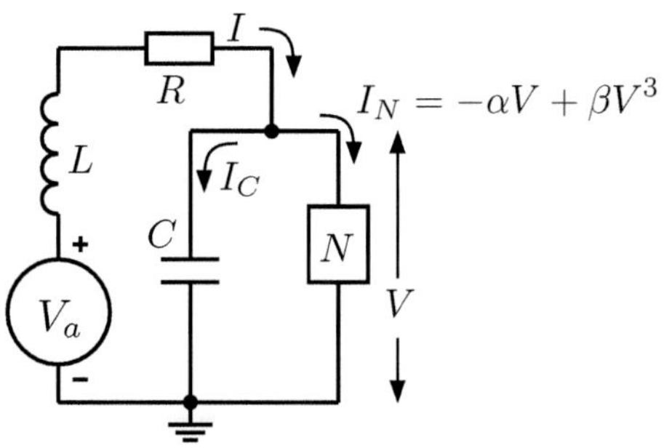

Fig. 5.17 Electronic circuit implementation of the FitzHugh-Nagumo model based on the RLC-circuit

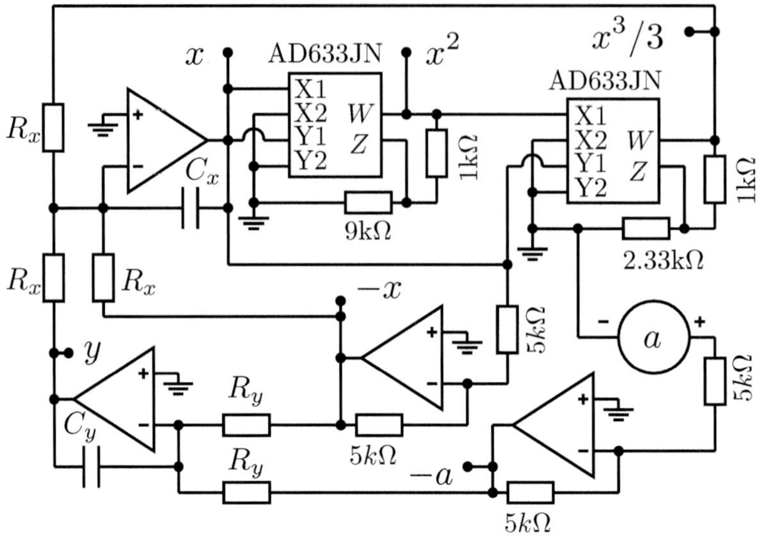

Fig. 5.18 Analog model developed for the FitzHugh-Nagumo oscillator (5.23), where $R_x = 1\,\text{k}\Omega$, $C_x = 5\,\text{nF}$, $R_y = 10\,\text{k}\Omega$, $C_y = 10\,\text{nF}$. Used operational amplifiers are TL072CP

$$\begin{cases} C\dfrac{dV}{dt} = I + \alpha V - \beta V^3, \\[2ex] L\dfrac{dI}{dt} = V_a - RI - V. \end{cases} \tag{5.27}$$

Introducing dimensionless variables $x = V/v_0$ Volt and $y = -I/i_0$ and dimensionless time $\tau = [v_0/(i_0 L)]t$ (here, $v_0 = 1$ Volt, $i_0 = 1$ Ampere), one can rewrite Eqs. (5.27) in the dimensionless form:

$$\begin{cases} \varepsilon\dfrac{dx}{d\tau} = \alpha x - \beta x^3 - y, \\[2ex] \dfrac{dy}{d\tau} = a + x - \gamma y, \end{cases} \tag{5.28}$$

where $\varepsilon = \dfrac{C}{L}(v_0/i_0)^2$, $\gamma = R/1\Omega$, $a = -V_a/v_0$. Model (5.28) is reduced to system (5.23) for $\alpha = 1$, $\beta = 1/3$, $\gamma = 0$.

The analog circuit in Fig. 5.18 represents the FitzHugh-Nagumo analog model proposed in Ref. [141]. The circuit equation are:

$$\begin{cases} R_x C_x \dfrac{dx}{dt} = -y + x - x^3/3, \\ \\ R_y C_y \dfrac{dy}{dt} = a + x, \end{cases} \tag{5.29}$$

System (5.29) is rewritten into form (5.23) with $\varepsilon = \dfrac{R_x C_x}{R_y C_y}$ by substituting the dimensionless time $\tau = t/(R_y C_y)$. If one needs to implement the system under an external perturbation, the external signals could be added into the circuit as additional input signals of the integrators connected through resistors.

5.4 Deterministic Chaos

A manifold of dynamical systems exhibiting the effect of deterministic chaos is incredible broad. Among them are famous examples of chaotic oscillators such as a model for atmospheric convection developed by Edward Lorenz more than fifty years ago [142], the Rössler oscillator proposed as a prototype equation to the Lorenz model [143], Chua's model [144, 145] and the Anishchenko-Astakhov self-oscillator [146, 147] describing electronic circuits, a model of a ring cavity containing a nonlinear dielectric medium proposed by Kensuke Ikeda [148, 149], the Mackey–Glass equations [150] used to model the variation in the relative quantity of mature cells in the blood. Nowadays, one can find a huge number of models developed for chaotic processes in neuroscience [151–153], chemistry and biochemistry [154, 155], ecology [156–158], population dynamics [159, 160], optics [161–164], electronics [165–168], plasma physics [169, 170], geology [171], economics [172–175], to name only a few.

In the simplest case, routes to chaos are realized in three-dimensional dynamical systems. In particular, one can observe three major routes to the chaotic dynamics in three-dimensional phase space: the period-doubling cascade route, the crisis and the intermittency route, and the route to chaos through quasiperiodicity destruction. In the presence of time delay mechanisms of the deterministic chaos occurrence can be more complicated since the time delay oscillators are in fact infinite-dimensional dynamical systems [176–178]. In addition, one can distinguish intrinsic peculiarities of bifurcation mechanisms resulting in complex chaotic dynamics such as noise-induced chaos [92, 179], the dynamics of chaotic Hamiltonian systems [180, 181] and systems characterised by the presence of hidden chaotic attractors in the phase space [182–184], spatio-temporal chaos accompanied by patten formation processes [158, 167, 185]. In the current section, we take into consideration only the route to chaos through the cascade of period-doubling bifurcations illustrated on examples of the Rössler and the Anishchenko-Astakhov self-oscillators.

5.4.1 Rössler Oscillator

The Rössler oscillator [143] is a well-known chaotic oscillator model exhibiting the chaotic dynamics associated with an attractor of the spiral type in the three-dimensional phase space. This attractor has certain similar properties as compared to the Lorenz attractor, but is simpler. Transition to chaos observed in the Rössler oscillator occurs via the cascade of the period-doubling bifurcations also called the Feigenbaum scenario. The Rössler oscillator equations are:

$$\begin{cases} \dfrac{dx}{dt} = -y - z, \\ \dfrac{dy}{dt} = x + \alpha y, \\ \dfrac{dz}{dt} = \beta + z(x - \mu). \end{cases} \tag{5.30}$$

To visualize transition to chaos in model (5.30), the parameters α and β are fixed, $\alpha = \beta = 0.2$, whereas the parameter μ varies in the range $\mu \in [2.0 : 5.0]$.

The evolution of system's attractor caused by increasing parameter μ is illustrated in Fig. 5.19. Initially, the system has a stable limit cycle corresponding to self-oscillations with a period T (Fig. 5.19a). As μ is increased, the supercritical period-doubling bifurcation takes place at μ_2^*, resulting in the birth of a stable limit cycle with a period $2T$ (Fig. 5.19b). After that, the system exhibits an infinite sequence of period-doubling bifurcations of cycles $2^k T$ (limit cycles with periods $4T$ and $8T$ are depicted in Fig. 5.19c, d).

When $k \to \infty$, the bifurcation points μ_k^* accumulate to some critical value $\mu = m_{\text{cr}}$, at which the period becomes infinite. At $\mu > \mu_{\text{cr}}$, there occur aperiodic oscillations being unstable according to Lyapunov. These oscillations follow a chaotic attractor in the system phase space Fig. 5.19e.

One can implement system (5.30) electronically by means of analog modelling. One of the possible circuit solution presented in paper [186] is depicted in Fig. 5.20. The circuit equations are:

$$\begin{cases} RC\dfrac{dx}{dt} = -y - z, \\ RC\dfrac{dy}{dt} = x + \alpha y, \\ RC\dfrac{dz}{dt} = \beta + z(x - \mu), \end{cases} \tag{5.31}$$

where $R = 10\,\text{k}\Omega$, $C = 10\,\text{nF}$. Here, the parameters β and μ are associated with DC-voltage sources, whereas the parameter α is determined by the ratio $\alpha = R_2/R_1$. It is important to note that the term being responsible for the third equation nonlinearity, $-z(x - \mu)/10$, is not amplified by 10 times before it becomes an input signal for the integrator. This circuit's property is realized to avoid the appearance of extremely high voltage amplitudes which could not be produced by operational

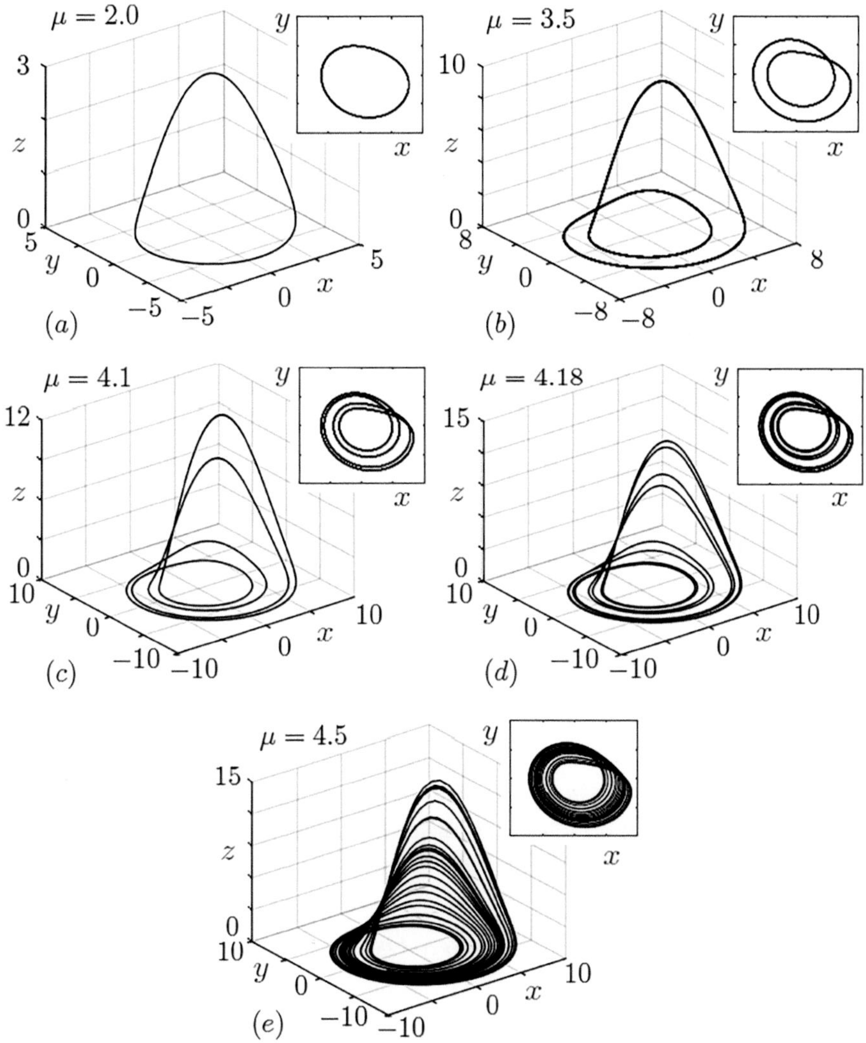

Fig. 5.19 The attractor evolution in the phase space of model (5.30): phase trajectories in the three-dimensional phase space and their projections on the plane (x,y). Other parameters are: $\alpha = \beta = 0.2$

amplifiers and analog multipliers (output voltage cannot exceed the voltage of the circuit power supply). The absence of multiplication by 10 is compensated by the presence of the integrator input resistor $0.1R = 1\,\mathrm{k\Omega}$. For this reason, the third equation of system (5.31) does not contain factor 1/10 at the nonlinear term. In addition, this approach allows to simplify the circuit and not to add one more non-inverting amplifier with the gain factor 10.

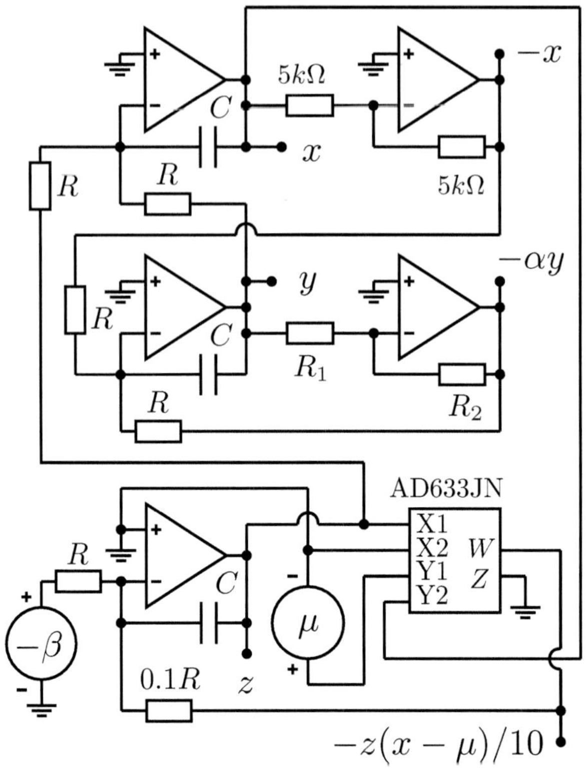

Fig. 5.20 Analog model of the Rössler oscillator (5.30), where $R = 10\,\text{k}\Omega$, $C = 10\,\text{nF}$, $\alpha = R_2/R_1$. Used operational amplifiers are LF412CN

5.4.2 Anishchenko-Astakhov Self-Oscillator

Changes exemplified in Fig. 5.19 happen in the similar way in the self-oscillator with inertial nonlinearity proposed by Vadim Anishchenko and Vladimir Astakhov [146, 147]. The Anishchenko-Astakhov self-oscillator illustrates the mechanism of birth and properties of a nonhyperbolic attractor in systems with a saddle-focus separatrix loop. This model is described by the following three-dimensional two-parameter differential system:

$$\begin{cases} \dfrac{dx}{dt} = mx + y - xz, \\ \dfrac{dy}{dt} = -x, \\ \dfrac{dz}{dt} = -g(z - \Phi(x)), \end{cases} \tag{5.32}$$

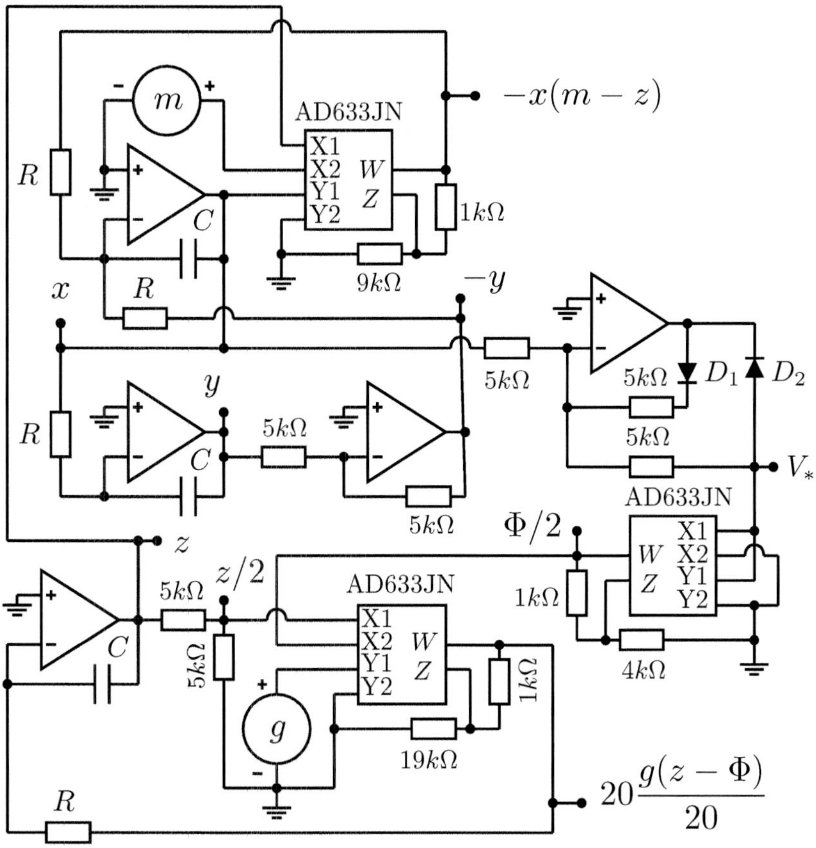

Fig. 5.21 Analog model of the Anishchenko-Astakhov self-oscillator (5.32), where $R = 10\,\mathrm{k\Omega}$, $C = 20\,\mathrm{nF}$. Used operational amplifiers are TL072CP. Elements D_1 and D_2 are diodes 1N4007

where $\Phi(x) = x^2$ for $x \geq 0$ and $\Phi(x) \equiv 0$ for $x < 0$. The route to chaos through a cascade of period doubling bifurcations occurs in the Anishchenko-Astakhov self-oscillator in similar way in comparison with the Rössler oscillator and the corresponding phase portraits are qualitatively identical to ones depicted in Fig. 5.19. For this reason, the attractor evolution in the phase space of the Anishchenko-Astakhov self-oscillator is not illustrated in the current book (the reader can find these illustrations in books [91, 187]).

An analog model of the Anishchenko-Astakhov self-oscillator is proposed in publications [109–111, 188] for studying stochastic Andronov-Hopf bifurcation by means of electronic experiments. The analog model circuit is depicted in Fig. 5.21. Its unique feature is the fact that the parameter values are defined by input voltage values. Due to this properties one can study the system under the impact of parametric modulation or parametric noise. The circuit equation are:

$$\begin{cases} RC\dfrac{dx}{dt} = x(m - z) + y, \\ RC\dfrac{dy}{dt} = -x, \\ RC\dfrac{dz}{dt} = -g(z - \Phi(x)), \\ \Phi(x) = \begin{cases} x^2, & x \geq 0, \\ 0, & x < 0, \end{cases} \end{cases} \tag{5.33}$$

where $R = 10\,\text{k}\Omega$, $C = 20\,\text{nF}$. As depicted in Fig. 5.21, the piecewise-smooth function $\Phi(x)$ is electronically implemented by using the op-amp-based inverting half-wave rectifier (see Sect. 3.2.15) whose output voltage is designated as V_*. The signal V_* takes the values $V_*(x) = -x$ for $x \geq 0$ and $V_*(x) \equiv 0$ for $x < 0$. Then this signal is processed by the analog multiplier and one obtains the signal $\Phi/2$ being equal to $\Phi/2 = V_*^2/2$.

5.5 Memristor

There are well-known three fundamental two-terminal passive electronic circuit elements: the resistor, the capacitor, and the inductor. The fourth element called the memristor was postulated by Leon Chua [189] in 1971. Chua's memristor relates the transferred electrical charge, $q(t)$, and the magnetic flux linkage, $\varphi(t)$: $d\varphi = M \cdot dq$, whence it follows that $M = M(q) = \dfrac{d\varphi}{dq}$. By using the formulas $d\varphi = Udt$ and $dq = idt$ (U is the voltage across the memristor, i is the current passing through the memristor) the memristor current-voltage characteristic can be derived: $U = M(q)i$. It means that M is the current-controlled resistance (memristance) and depends on the entire past history of $i(t)$:

$$M(q) = \frac{d\varphi}{dq} = \varphi'\left(\int_{-\infty}^{t} i(t)dt\right). \tag{5.34}$$

In a similar way, it can be written $dq = W \cdot d\varphi$, and then $W = W(\varphi) = \dfrac{dq}{d\varphi}$ and $i = W(\varphi)U$. Therefore W is the flux-controlled conductance (memductance) and is the function of the entire past history of $U(t)$:

$$W(\varphi) = \frac{dq}{d\varphi} = q'\left(\int_{-\infty}^{t} U(t)dt\right). \tag{5.35}$$

The Eq. (5.34) describes the charge-controlled memristor, and the Eq. (5.35) describes the flux-controlled one. Two forms (5.34) and (5.35) are mutually equivalent. Choice of the function describing the memristor depends on specific of each considered problem. The fundamental memristor property becomes highlighted from (5.34) and (5.35). It is the dependence of electrical characteristics on past history of functioning. Initially, the memristor was introduced as realization of a hypothesis of relationship between the electrical charge and the magnetic flux linkage. Then the term "memristor" was extended to the concept of "memristive systems". According to the paper [190] a class of the memristive systems is identified by continuous functional dependence of characteristics at any time on previous states of the memristive system, and is described by the following equations:

$$y = g(z, x, t)x, \quad \dot{z} = f(z, x, t), \tag{5.36}$$

where x is the input signal, y is the response of the system, the vector $z \in \mathbb{R}^n$ denotes the system state, $f(z, x, t)$ is a continuous n-dimensional vector function, $g(z, x, t)$ is a continuous scalar function. It is assumed that the state equation $\dot{z} = f(z, x, t)$ has a unique solution for any initial state $z_0 \in \mathbb{R}^n$. It is important to note that such definition is mathematical interpretation, which does not concern a physical sense of the dynamical variables and their functional dependence. L. Chua and co-authors marked the fact that the Eqs. (5.36) do not impose limitations on nature of the memristive systems [191, 192]. For example, models of an electronic element thermistor and a discharge tube satisfy the definition (5.36) of the memristive system [190, 193]. There are well-known examples of the memristive systems, which were developed more than a century ago [194]. As will be shown below, the memristive systems are of a frequent occurrence in different fields of physics as well as in neurodynamics and other sciences concerning biological systems.

At the moment of publication the first Chua's papers concerning the memristor were less famous than now. A flurry of interest arose when scientists from the Hewlett Packard Company announced fabrication of TiO_2 oxides-based electronic two-terminal device, which exhibits features of the memristor dynamics [195]. The authors have developed a mathematical model of the experimental prototype, which satisfies the definition (5.36). The resistance of the experimental model presented in [195] depends on quantity of electricity (charge) passed through the device. In this context, the created passive two-terminal element can be assumed to be experimental realization of the memristor. Nevertheless, experimental demonstration did not close the question of the memristor existence, since a magnetic field is not involved in experimental prototype functioning. It should be pointed out, that the described in [195] effects were highlighted during the late 1960s and called the "resistive switchings" [196].

At the present time many objects, which exhibit the memristor dynamics, are known. Their characteristics reflect information about pre-history of operation. They are called the memristors in the following. Some of them are the oxide-based memristors [195–207], the polymer memristors [208–212], the ferroelectric memristors [213, 214], the memristors, based on resonant tunneling diodes [215–217], the spintronic memristors, operation principles of whose are based on magnetic phenomena [218–220]. Certain electronic circuits of one-port devices are realization of the memristor [221, 222].

The memristors have a wide range of practical application. First of all, they are considered as a memory element. Memristor peculiarities can be used for analog circuit development, such like programmable amplifiers and attenuators, adaptive filters, to name only a few. The memristor attracts worldwide attention due to its potential applications in construction of next generation computers [223–226]. A memristor potential for using in electronics is shown in the papers [227–231] and in the books [232–236] in details.

Appearance of the memristors has given an impulse to progression in certain fields of neurodynamics [235, 237]. Particularly, an analogy between the memristor dynamics and the behavior of a neural cell synapse [238–242] makes be possible to consider the memristor as an element of neuromorphic electronic system working on principles of brain functioning [243–245]. An interesting example of that similarity has been mentioned in the paper [246]. The authors have revealed the ability of memristor-based circuits to adapt to external signals and compared the obtained results with Pavlov's experiments in conditioned and unconditioned reflexes formation.

Now the issues addressing the memristor remain to be in progress. Some recently obtained results are collected in books [247, 248], which include the issues of modelling and theoretical aspects as well as circuit implementations and practical applications. Novel papers provide an encouraging pathway toward neuromorphic computing [249–252], particularly in fuzzy logic systems [253].

In the context of nonlinear dynamics, the memristor attracts an interest as an element, intrinsic properties of whose can essentially change the dynamics of electronic oscillatory systems and are responsible for qualitatively new types of the behavior. There are examples of memristor-based chaotic oscillators [254–258] and Hamiltonian systems including the memristor [259]. There is a variety of publications addressing the issue of the collective dynamics of coupled through the memristor oscillators and ensembles of coupled memristor-based oscillators: from synchronization of two chaotic or regular self-sustained oscillators coupled through the memristor [260–267] to spatio-temporal phenomena in ensembles of interacting oscillators with adaptive coupling realized by the memristor or in ensembles of coupled memristor-based oscillators [268–277]. Next interesting effect is stochastic resonance, which has been revealed in the single memristor under the summary influence of noise and periodic signal [278]. A broad variety of the effects is complemented by the existence of hidden attractors in models including memristor [279–281].

All the mentioned above and other oscillatory regimes observed in memristor-based oscillators can be experimentally studied and applied on examples of their electronic implementations. For this purpose, one can use passive memristive two-terminal elements or electronic circuits emulating the memristor. The second approach is discussed in the further section.

5.5.1 *Transistor-Switch-Based Implementation of Chua's Memristor*

One of the simplest memristor model is Chua's memristor described by the piecewise-linear dependence $q(\varphi)$ which takes the following form for the flux-controlled memristor:

$$q(\varphi) = \begin{cases} (a-b)\varphi_* + b\varphi, & \varphi \geq \varphi_*, \\ a\varphi, & |\varphi| < \varphi_*, \\ -(a-b)\varphi_* + b\varphi, & \varphi \leq -\varphi_*, \end{cases} \tag{5.37}$$

where φ plays a role of the memristor state variable. Then the memristor conductance $W(\varphi)$ becomes:

$$W(\varphi) = \begin{cases} a, & |\varphi| < \varphi_*, \\ b, & |\varphi| \geq \varphi_*. \end{cases} \tag{5.38}$$

Nonlinearity (5.38) can be approximated by the hyperbolic tangent function:

$$W(\varphi) = \frac{b-a}{2}\tanh\left(k(\varphi^2 - \varphi_*)\right) + \frac{b+a}{2}, \tag{5.39}$$

where a parameter k characterizes the sharpness of the transitions between two memristor's states. It has been shown in [282] that changing the memristor conductance function to the smooth one does not qualitatively modify the memristor properties. The classical loop in the current-voltage characteristic of the memristor driven by the external periodic influence persists (see Fig. 2 in Ref. [282]). The memristor model including tanh-nonlinearity is not the only smooth memristor model. There is a number of smooth models describing various memristor properties [192, 234, 283–287].

Real memristive systems can 'forget' the state history over time, which results in finite correlation between current and previous states. That means the impact of the past states weakens with increase of the time distance between present and past states. For sufficiently long time distances the previous state impact can be neglected and one can assume that a memristive system forgot that states. Such effects in memristors based on metal oxides are associated with the diffusion of charged particles [286, 288, 289] (however, the 'forgetting' can happen very slowly). One of the simplest form of the memristor state equation which implies the forgetting effect is the following:

$$\frac{dz}{dt} = g(x, z) = x - \delta z, \tag{5.40}$$

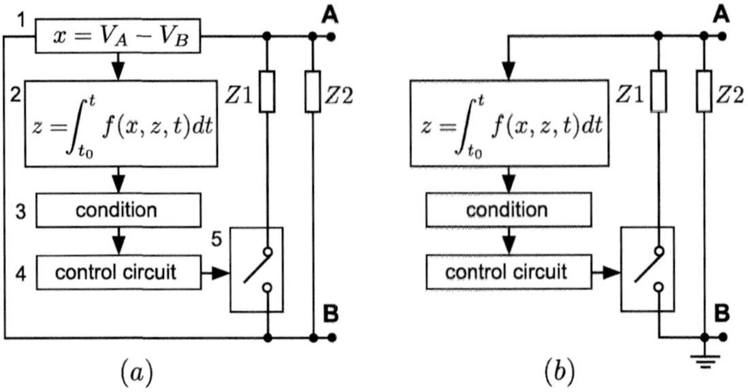

Fig. 5.22 Generalized scheme for memristive circuit synthesis. The difference between panels (**a**) and (**b**) consists in the presence of the reference point at terminal B

where z plays a role of a memristor state variable, x is an input signal, a parameter δ characterizes the forgetting effect strength.

The memristors well-described by piecewise-smooth functions or by continuous ones such as tanh-function can be synthesised as semiconductor devices as well as by means of analog modelling methods. The second approach implies the development of an analog one-port circuit with memristive properties. One of the possible ways to create memristive circuits implies using transistor-based switches. For example, such circuit is illustrated and described in paper [290]. The operation principle of the considered circuit is illustrated in Fig. 5.22a. It includes two parallel one-ports Z1 and Z2. Element Z2 is always connected with terminals A and B. The connection of element Z1 with terminal B depends solely on switch circuit 5 controlled by feedback. The feedback consists of blocks 1-4. Block 1 produces a signal being equal to the voltage drop across the load (between terminals A and B): $x(t) = V_A(t) - V_B(t)$. If terminal B is connected with the ground point, block 1 can be excluded as in Fig. 5.22b. Then $x(t)$ is the input signal for integrator 2, which forms the signal $z(t)$ according to the state equation of system (5.36). After that, the signal $z(t)$ reaches the block of condition (block 3). If the condition is fulfilled, the corresponding signal comes to control circuit 4, which opens and closes switch 5. When switch 5 is closed, the circuit conductance is the sum of the conductances of elements Z1 and Z2. Otherwise, the switch is open and the conductance of element Z1 is excluded from the resulting circuit conductance. It is assumed that the controlling feedback has zero input current and has no time delay.

The approach for synthesis of memristive one-ports with piecewise-smooth characteristics proposed in [290] is testified on an example of the current-voltage characteristic of circuits driven by external periodic forcing (see Fig. 5.23a). The dimensionless model for Chua's memristor driven by a periodic signal (the voltage V_e applied across the memristor) takes the form

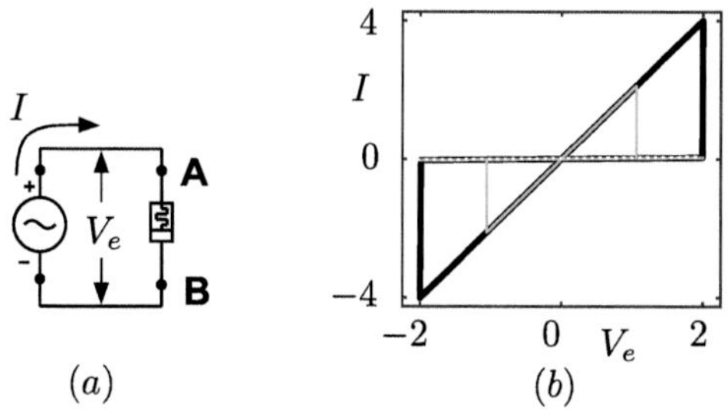

Fig. 5.23 Piecewise-smooth memristor model under periodic forcing (5.41): circuit model (panel (**a**)) and its current-voltage characteristic (panel (**b**)) for increasing frequency of external periodic forcing: $f_e = 0.32$ (black solid curve), $f_e = 0.59$ (gray solid curve), $f_e = 0.65$ (white dotted curve). Other parameters are: $a = 0.02$, $b = 2$, $\varphi_0 = 1$, $V_{\max} = 2$

$$I = G_M(\varphi)V_e, \quad \frac{d\varphi}{dt} = V_e, \quad G_M(\varphi) = \begin{cases} a, & |\varphi| \le \varphi_0, \\ b, & |\varphi| > \varphi_0, \end{cases} \tag{5.41}$$

where I is the dimensionless current passing through the memristor, V_e is the dimensionless voltage across the memristor, φ is the dimensionless flux controlling the memristor state, the dimensionless parameter $G_M(\varphi)$ plays a role of the flux-controlled memrsitive conductance. In the presence of the external periodic signal $V_e(t) = V_{\max}\sin(2\pi f_e t)$ (see Fig. 5.23a), memristor model (5.41) exhibits the current-voltage characteristic shaped as the classical pinched hysteresis loop (Fig. 5.23b). When the frequency f_e increases, the pinched hysteresis loop transforms into a single-valued function.

The electronic circuit realization of system (5.41) depicted in Fig. 5.24a is developed according to the principles described above. The circuit has two terminals A and B. Element Z2 is a resistor ($R_2 = 50\,\text{k}\Omega$) which is always connected with terminals A and B. The second load element, Z1, is a resistor ($R_1 = 1\,\text{k}\Omega$) connected between terminals A and B in parallel to R_2 when the electronic switch is closed. In this case the total resistance between terminals A and B is $R_{\text{closed}} = \frac{R_1 \times R_2}{R_1 + R_2} \approx 980.4\,\Omega$. When the electronic switch is open, the total resistance becomes $R_{\text{open}} = R_2 = 50\,\text{k}\Omega$. The electronic AC switch controlled by feedback consists of two transistors Q1 and Q2 connected with diodes D3, D4 and resistors. The feedback consists of several blocks. The first element is analog follower A1 which output signal is denoted as $V_e = V_e(t)$. Then the output signal of inverter A2 equals to $-V_e(t)$. The signal $-V_e$ comes to analog integrator A3, which is responsible for the differential equation defining the state variable z. The equation for integrator A3 is $RC\frac{dz}{dt} = V_e - \frac{10\,\text{k}\Omega}{50\,\text{k}\Omega}z$, where $z(t)$ is the integrator output voltage. Then the condition denoting the piecewise-smooth characteristic is formed. Operational amplifiers A4 and A5 extract the absolute value of the $z(t)$ variable (the voltage at operational amplifier A5 output equals to $-|z(t)|$). Operational amplifier A6 has the summing amplifier configuration with the output voltage being equal to $|z(t)| - 1$. This signal comes to comparator A7. If the signal $|z(t)| - 1$ is greater than zero

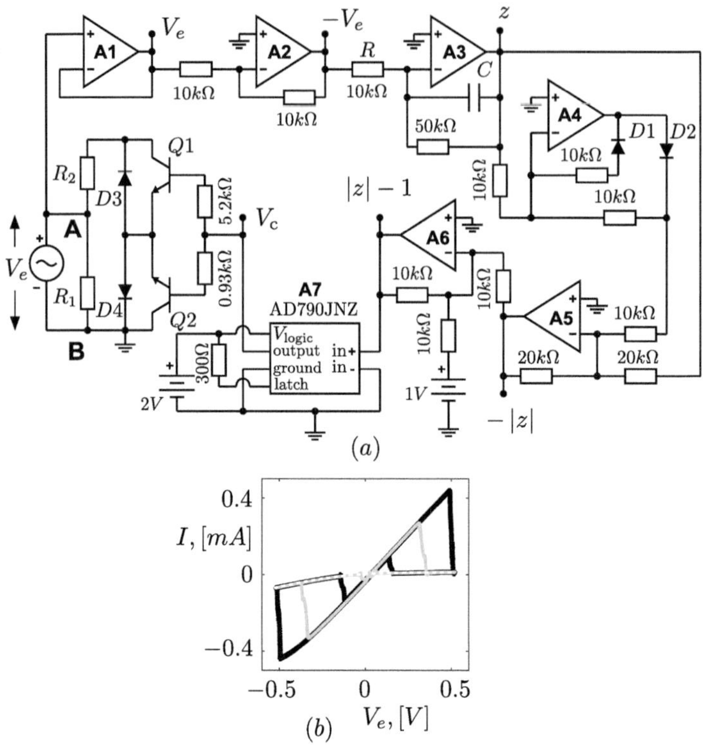

Fig. 5.24 (**a**) Circuit diagram of the particular memristor circuit implementation (see Eq. (5.42)). Elements D1 and D2 are diodes 1N4148, D3 and D4 are diodes 1N4007, Q1 and Q2 are transistors 2N2222A, A1-A6 are operational amplifiers TL072CP; (**b**) Experimentally acquired characteristic of the experimental setup (5.42) under external driving $V_e(t) = V_{max}\sin(2\pi f_e t)$. The parameters are: $V_{max} = 0.5$ V, $f_e = 4.0$ kHz (black solid line), $f_e = 7.0$ kHz (gray solid line), $f_e = 8.1$ kHz (white dotted line)

(it corresponds to condition $|z(t)| > 1$), the corresponding signal is generated at the output of comparator A7, and the transistor-based switch becomes closed. The equations describing the circuit are:

$$I = G_M(z)V_e, \quad RC\frac{dz}{dt} = V_e - 0.2z,$$
$$G_M(z) = \begin{cases} \dfrac{1}{R_2}, & |z| \le 1\text{ V}, \\[2ex] \dfrac{1}{R_1} + \dfrac{1}{R_2}, & |z| > 1\text{ V}. \end{cases} \tag{5.42}$$

It is important to explain a role of the term $-0.2z$ in the state equation of system (5.42). Let us assume that the term $-0.2z$ is absent. In such a case, the expression for the state variable takes the form $z(t) = \frac{1}{RC} \int_0^t V_e(t)dt$. When the circuit is driven by a periodic influence $V_e(t)$, then the voltage across the load is $V(t) = V_e(t) = V_{max} \sin(2\pi f_e t)$. The variable z is expressed as $z(t) = -\frac{1}{RC2\pi f_e} \cos(2\pi f_e t) + K$ in the absence of the term $-0.2z$ in the state equation. It represents periodic oscillations with fixed mean value K. However, the behaviour observed in electronic experiments differs from the expected dynamics. Since the initial power-up moment, the variable z increases (or decreases) and reaches its maximal (minimal) value limited by the power-supply voltage. This is due to the following facts. The equation for OA-based integrator does not take into consideration certain peculiarities of real operational amplifiers, such as a small but nonzero input current and capacitor's leakage currents. Moreover, the external periodic signal can slightly fluctuate or have a nonzero mean value. In order to stabilize the oscillations of the variable z and make the dynamics less sensitive to random perturbations, resistor R4 being responsible for the term $-0.2z$ was introduced into the circuit. In addition, $-0.2z$ plays a role of the term being responsible for the memristor forgetting effect.

Experimentally registered current-voltage characteristic (Fig. 5.24b) demonstrates all the phenomena observed in the numerical model. The quantitative difference consists in the frequency range, where the memristive peculiarities are exhibited. In practice, this range is defined by the ratings of R and C.

The principal difference between the discussed memristor implementation approach and a manifold of already published solutions (for instance, see the circuit implementations for the cubic memristor [221] and memcapacitor [291], the diode-bridge-based solution [222] and the analog emulator of the memristor with tanh-nonlinearity [292] consists in the prospects for the piecewise-smooth nonlinearity realization.

The advantage of the proposed circuit is its universality. Indeed, adjusting the circuit component ratings, one can control a frequency range, where the memristive properties are exhibited, as well as the resulting circuit impedance in different regimes. The proposed circuit solution allows for the implementation of any deterministic or stochastic forms of the memristor state equation. For instance, one can take into account the memristor forgetting effect [286, 288, 289] or diffusive processes in the memristor [293] and experimentally study the manifestations of such phenomena. Moreover, when the switch is open or closed, the reactive properties of the load can be principally different (for example, elements Z1 and Z2 can be capacitors, resistors, etc.). Thus, the described circuit can be applied for a broad variety of applications, including synthesis of memristive oscillators, adaptive filters, amplifiers and attenuators, etc.

5.5.2 Analog Model of Chua's Memristor

The second approach for the memristive circuit implementation is to realize memristive blocks as an analog model where all the signals and dynamical variables are voltages (in contrast, the experimental realization of memristor in the previous section is characterised by the voltage across the memristive circuit and the current passing through it). One of the possible solution for such kind of memristor modelling is proposed in paper [294]. The discussed circuit implementation is depicted in Fig. 5.25. Here, the comparator is configured such that the memristive function G_M takes values $G_M = 0.02$ for $|z| \leq 1$ and $G_M = 2$ for $|z| > 1$. Then the equation for the output signal is:

$$V_{\text{out}} = xG_M(z), \quad RC\frac{dz}{dt} = x,$$
$$G_M(z) = \begin{cases} 0.02, & |z| \leq 1, \\ 2, & |z| > 1. \end{cases} \tag{5.43}$$

Similarly to the circuit solution in Fig. 5.24, the analog memristor model is quite flexible: one can vary R and C to control the frequency range, where the memristive properties are manifested. In addition, one can implement any deterministic or stochastic forms of the memristor state equation.

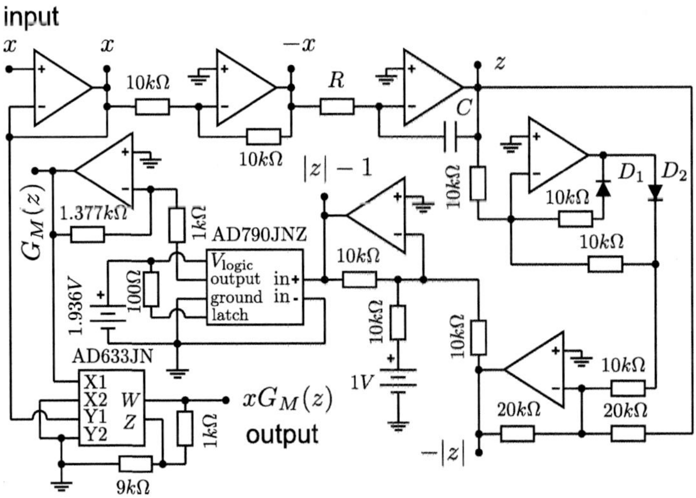

Fig. 5.25 Analog model of Chua's memristor (Eq. (5.41)) described by Eq. (5.43). Elements D1 and D2 are diodes 1N4007. The used operational amplifiers are TL072CP

5.5.3 Memristor-Based Oscillators with a Line of Equilibria

The simultaneous and continuous dependence of the oscillatory dynamics both on parameter values and initial conditions is a frequent occurrence in various dynamical systems. In most cases, such peculiarities are associated with the properties of conservative oscillators. However, such behaviour can be exhibited by nonlinear dissipative oscillators with manifolds of equilibria. The manifolds of equilibria consist of non-isolated equilibrium points and takes different forms: line of equilibria [295–298], surface of equilibria [299], circle [300, 301] and square [301] equilibria, etc. A special class of such manifolds is represented by m-dimensional normally hyperbolic manifolds of equilibria characterized by m pure imaginary eigenvalues or m eigenvalues being equal to zero, whereas all the other eigenvalues have non-zero real parts. In the simplest case $m = 1$, these manifolds constitute a line of equilibria. The significant feature of systems with a line of equilibria is the occurrence of so-called bifurcations without parameters, i.e., the bifurcations corresponding to fixed parameters when the normal hyperbolicity condition is violated at some points of the line of equilibria [295–298, 302–304]. Resultantly, the bifurcations without parameters can be observed by means numerical simulations of dynamical systems starting from varying initial conditions.

Memristor-based oscillators are widely represented in a variety of systems with a line of equilibria. One of the frequently considered models is the series RLC-circuit with negative resistance, where a flux-controlled memristor is connected in parallel with the capacitor. Bifurcation mechanisms of the periodic solution appearance in this system have been explored numerically and analytically for different kinds of nonlinearity [282, 294, 302, 305–308]. Analytical solution of the model equations has enabled to recognize significant features of the supercritical [307, 308] and subcritical [282] Andronov-Hopf bifurcation. In addition, the recently published research [282] has demonstrated how the saddle-node bifurcation of limit cycles observed in classical self-oscillators with hard self-oscillation excitation transforms in systems with a line of equilibria. Thus, all the bifurcations of limit cycles exhibited by self-oscillators with one degree of freedom have been described in terms of oscillators with a line of equilibria. The similarity between classical self-oscillators and the considered systems with a line of equilibria is emphasized by the fact that undamped periodic oscillations in systems with a line of equilibria can be synchronized by external periodic forcing [309].

The period-doubling bifurcations considered in the recent research [310] are induced by changing the initial conditions whereas parameter values are fixed. These bifurcations were revealed in a modified Anishchenko-Astakhov self-oscillator where the ability to exhibit bifurcations without parameters is associated with the properties of a memristor. Thus, the studied bifurcations can be classified as the period-doubling bifurcations without parameters. Moreover, the demonstrated in Ref. [310] cascade of the period-doubling bifurcations without parameters results in transition to deterministic chaos called 'Feigenbaim scenario without parameters'.

The circuit model proposed in paper [311] represents a universal dynamical system for the implementation of bifurcations without parameters related to basic bifurcations of steady states: the pitchfork, transcritical and saddle-node bifurcations. The demonstrated bifurcations are associated with the existence of lines of equilibria. The transcritical and saddle-node bifurcations imply the coexistence of a stable and unstable line of equilibria. The pitchfork bifurcation realises the transition to the regime of bistability which consists in the coexistence of two lines of equilibria attracting the phase trajectories and an unstable line of equilibria between them.

A spectrum of bifurcations without parameters mentioned above are analogues of classical bifurcations of limit cycles (the Andronov-Hopf bifurcation, the saddle-node bifurcation, the period-doubling bifurcation) and steady states (the pitchfork bifurcation, the saddle-node bifurcation, the transcritical bifurcation) occurring in oscillators with a finite number of isolated limit sets. However, it is extremely difficult to implement bifurcations without parameters in physical experiments. In particular, this fact was emphasized in Ref. [294] on an example of a memristor-based oscillator:

$$\frac{dx}{dt} = y - G_M(z)x,\ \frac{dy}{dt} = -x + \beta y,\ \frac{dz}{dt} = x, \\ G_M(z) = \begin{cases} a, & |z| \le 1, \\ b, & |z| > 1, \end{cases} \tag{5.44}$$

where the memristor parameters are $a = 0.02$, $b = 2$. System (5.44) has a line of equilibria, i.e., each point on the axis OZ (the points with coordinates $x = y = 0$, $z \in (-\infty : \infty)$) is an equilibrium point. One of the eigenvalues ρ_i of the equilibria is always equal to zero and the others depend on the parameters and the position of the point on the OZ-axis (z coordinate):

$$\rho_1 = 0,\ \rho_{2,3} = -\frac{\alpha G_M(z) - \beta}{2} \pm \sqrt{\frac{(\alpha G_M(z) - \beta)^2}{4} - \alpha\beta G_M(z) - \alpha\gamma}. \tag{5.45}$$

Each point of the line of equilibria is neutrally stable in the OZ-axis direction. At the same time, for each point of the line of equilibria one can distinguish the plane $Q(x, y, z)$, which includes trajectories corresponding to repelling or attraction of this point in its vicinity. In the plane $Q(x, y, z)$ stability of the equilibrium point belonging to the line of equilibria is cracterized by means of linear stability analysis used for isolated fixed points on the phase plane. Hereinafter, using the terms "stable" or "unstable" point at the line of equilibria, we mean the behavior of the trajectories in the neighbour of the equilibrium point in the plane $Q(x, y, z)$.

Increasing the parameter β from zero gives rise to the following bifurcational changes in the phase space. In case $0 \le \beta < 0.02$, an attractor of the system is a manifold of stable equilibria, and all trajectories are attracted to them (see Fig. 5.26a). When parameter β reaches the value $\beta = 0.02$, equilibrium points with

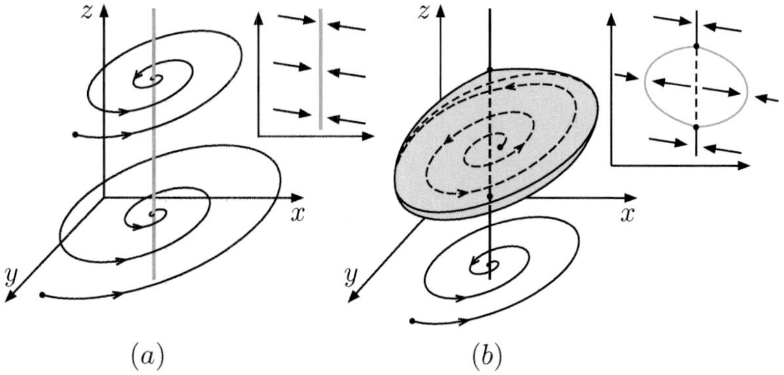

Fig. 5.26 Illustration of self-oscillation excitation in three-dimensional systems with a line of equilibria associated with the supercritical Andronov-Hopf bifurcation. Before the bifurcation all the trajectories are attracted to a line of equilibria (panel (**a**)). After the bifurcation all the trajectories are attracted to either stable semi-axes on the line of equilibria or invariant closed curves on a two dimensional surface (panel (**b**)). The right insets schematically show a section of the attractor by any plane being parallel to the axis Oz and containing the line of equilibria (attraction is schematically shown)

coordinate $-1 < z < 1$ become unstable, while other points of the line of equilibria with $|z| \geq 1$ remain to be stable (they become unstable at $\beta = 2$). Starting from the vicinity of unstable point $(0; 0; z \in (-1; 1))$ the phase point moves away from the initial state and traces a spiral-like trajectory. That movement culminates in motion along an invariant closed curve (the black solid curve in Fig. 5.26b). Any changes of the initial conditions give rise to hit on another invariant closed curve. In that way, an attractor structure changes in the point $\beta = 0.02$. After the bifurcation the attractor consists of a continuous set of closed curves which form a two-dimensional surface (like a whirligig, is coloured in grey in Fig. 5.26b) for $-1 < z < 1$ and of a set of fixed points. Further increasing the parameter β up to $\beta = 2$ does not lead to qualitative and quantitative changes of the attractor. Thus, for $0.02 < \beta < 2$ one can realize a bifurcation without parameter by tuning the initial condition $z_0 = z(t = 0)$. It occurs at $z_0 = \pm 1$ where the observed oscillatory dynamics changes from the attraction to a line of equilibria to undamped periodic oscillations. In terms of phase portraits on the plane (x,y), this transition is very similar to the Andronov-Hopf bifurcation observed in classical oscillators with a finite number of isolated limit sets.

Experimental studies of model (5.44) were carried out in Ref. [294, 307] on an example of the corresponding analog model depicted in Fig. 5.27 where the third equation and the memristive piecewise-smooth nonlinearity is realized as in Fig. 5.25. The analog model equations are:

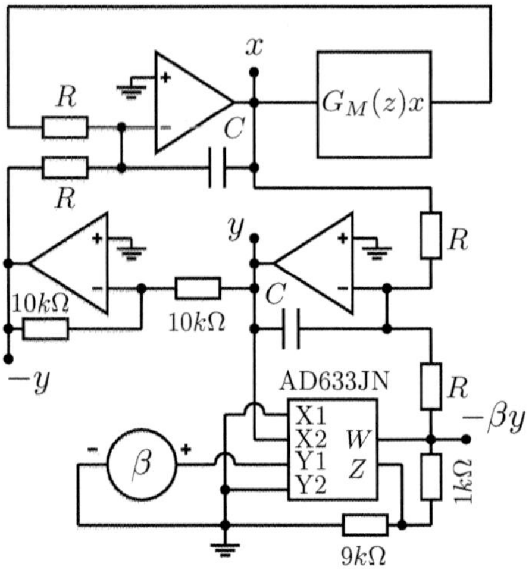

Fig. 5.27 Analog model of memristor-based oscillator (5.44), where $R = 56\,\text{k}\Omega$, $C = 100\,\text{nF}$. Used operational amplifiers are TL072CP. The circuit includes block $G_M x$ depicted in Fig. 5.25

$$
\begin{aligned}
&RC\frac{dx}{dt} = y - G_M(z)x, \; RC\frac{dy}{dt} = -x + \beta y, \; RC\frac{dz}{dt} = x, \\
&G_M(z) = \begin{cases} a, & |z| \le 1, \\ b, & |z| > 1, \end{cases}
\end{aligned}
\tag{5.46}
$$

where $R = 56\,\text{k}\Omega$, $C = 100\,\text{nF}$.

5.6 Dissipative Solitons in a Single Delayed-Feedback Oscillator

As was first mentioned in Ref. [312], there exists an analogy between the behavior of time-delayed systems and the dynamics of ensembles of coupled oscillators or spatially extended systems [313]. The similarity takes place when the delay time, τ, is much longer than system's response time, which allows the system to reveal spatio-temporal phenomena (for example, coarsening [314], chimera states [57, 315, 316], soliton dynamics [317–319]) in the purely temporal dynamics of a time-delay system. This space-time analogy can be obtained by implementing a space-time transformation of the delayed-feedback system, where the temporal dynamics is mapped onto space-time (σ, n) by introducing the space-time map $t = n\tau + \sigma$ with an integer (slow) time variable n, and a pseudo-space variable $\sigma \in [0, \eta]$, where $\eta = \tau + \delta$ with a small quantity δ. For each set of parameters a unique value η can be chosen such that the oscillatory dynamics is periodic with the period η. The mentioned technique is described in details in review [320] as well as a manifold of spatio-temporal phenomena tracked down in the dynamics of delayed-feedback oscillators. One of these effects is formation of dissipative solitons discussed in this section.

Dissipative solitons are stable localized structures, which are realized in nonlinear dissipative spatially extended systems [321–327]. Appearance of the dissipative solitons results from the energy supply-dissipation balance and the simultaneous impact of nonlinearity and dispersion. They occur in optics and optoelectronics [324, 328–334], magnetoelectronics [335, 336], plasma physics [337–339], biology [340, 341] and chemistry [342, 343]. In optics these structures are now under intensive study due to perspectives of practical applications in optical data processing and communication [344, 345] in the context of development of new generation computing systems [346] such as neuromorphic and reservoir computers [347, 348].

Examples of dissipative soliton patterns for delayed-feedback oscillators, in both theory and experiments, were reported recently in optical systems [317, 320, 324, 349]. Mathematical aspects of the dissipative soliton appearance were derived for a reduced time-delayed model of mode-locked laser [332, 350, 351], the complex Ginzburg-Landau equation including feedback terms [332, 333, 352], and the FitzHugh-Nagumo model with time-delay [349, 353].

The delayed-feedback oscillator proposed in paper [318] exhibits single as well as multiple dissipative soliton patterns clearly illuminated in the respective virtual space, which is caused by bistability of the nonlinear function applied through delayed feedback. The considered model equations are:

$$\begin{cases} \varepsilon\dot{x} = -y - gx + f[x(t-\tau)], \\ \dot{y} = x + my \end{cases} \tag{5.47}$$

with bimodal negative delayed-feedback function f chosen, for definiteness, as

$$f[x] = -\frac{x}{ax^2 + b}, \tag{5.48}$$

where small parameter $\varepsilon > 0$ separates slow and fast motions. Other parameters are $g, a, b, m > 0$.

The delay-free dynamics of the model (i.e. when $\tau = 0$ in Eq. (5.47)) is bistable and exhibits the coexistence of two self-oscillatory regimes [318]. When the delay τ is introduced and it is much larger than the characteristic response time and much smaller than the time of observation, the system exhibits periodically alternating impulses, *dissipative solitons* in some range of the parameters. Following the standard procedure [312, 313], delayed-feedback system (5.47) can be analyzed in a virtual, spatially extended representation. The resulting space-time plot corresponding to one impulse consists of a single localized perturbation surrounded by a plateaus of the almost constant amplitude (given by the quiescent regime) (Fig. 5.28a, b). This structure is topologically equivalent to dissipative solitons observed in spatially extended systems. The soliton shown in Fig. 5.28a corresponds to sharp decreasing the signal $y(t)$ while the soliton depicted in Fig. 5.28b represents increasing the instantaneous value $y(t)$. Hence, one can distinguish dark (Fig 5.28a)

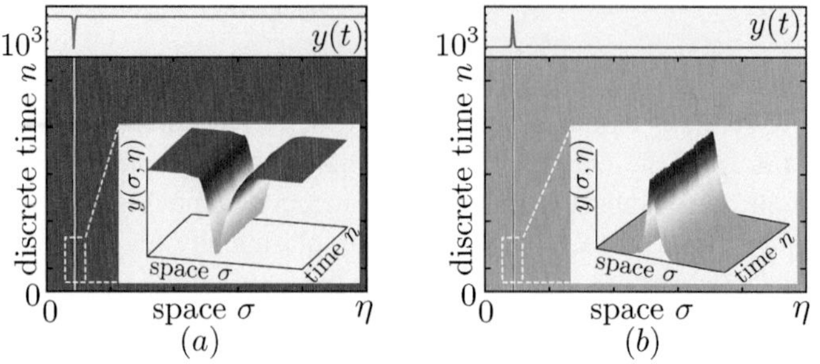

Fig. 5.28 Model (5.47): space-time plots of $y(t)$ corresponding to one dark soliton (panel (**a**)) and one bright soliton (panel (**b**)). Parameters are: $\varepsilon = 0.005$, $a = 200$, $b = 0.2$, $g = 0.1$, $m = 8.05$, $\tau = 50$. Quasi-space parameter η sets to 50.2333

and bright (Fig. 5.28b) solitons respectively [354]. As a result of the system symmetry, two soliton types—dark (spiking down Fig. 5.28a) and bright (spiking up, see Fig. 5.28b)—naturally arise as symmetric copies of each other with equal probability in a case of random initial conditions.

Soliton structures which were observed in numerical simulations of the model (5.47) have been also obtained in electronic experiment. In contrast to numerical simulations performed for the completely symmetric case, where dark or bright soliton structures have equal probability, the experimental implementation of bright solitons appears to be much more difficult. These structures are very rare in the experiment. Such dissimilarity occurs because of some intrinsic features, imperfections and asymmetry of the setup.

Physical experiment was carried out with using an experimental prototype being an electronic model of the system (5.47) implemented by principles of analog modelling. Detailed circuit diagram of the setup is shown in Fig. 5.29a. It contains two integrators, whose output voltages are taken as the dynamical variables, x and $10y$, respectively. The experimental facility includes a time-delay line, which was realized using personal computer complemented by an acquisition boards by National Instruments. The time delay value is constant, $\tau = 47$ ms. Equations describing operation of the experimental setup are the following:

$$\begin{cases} 0.1R_xC\dfrac{dx}{dt} = -y - gx + \dfrac{1}{20}f[x_\tau], \\ R_yC\dfrac{dy}{dt} = x + my, \end{cases} \tag{5.49}$$

where $x_\tau = x(t-\tau)$, $g = 0.1$, $m = R_m/1\,\text{k}\Omega$, $C = 50$ nF, $R_x = 1$ KΩ, $R_y = 20$ KΩ. Equations (5.49) include the function $f[x_\tau]$ depending on delayed signal x_τ and being responsible for delayed feedback. The function $f[x_\tau]$ was realized by using a block of nonlinear transformation (the right part of the circuit depicted in Fig. 5.29). The experimentally measured dependence $f[x_\tau]$ can be approximated by the formula $f[x_\tau] = -x_\tau/(a_e x_\tau^2 + b_e)$ with $a_e = 10.92$ and $b_e = 0.38$.

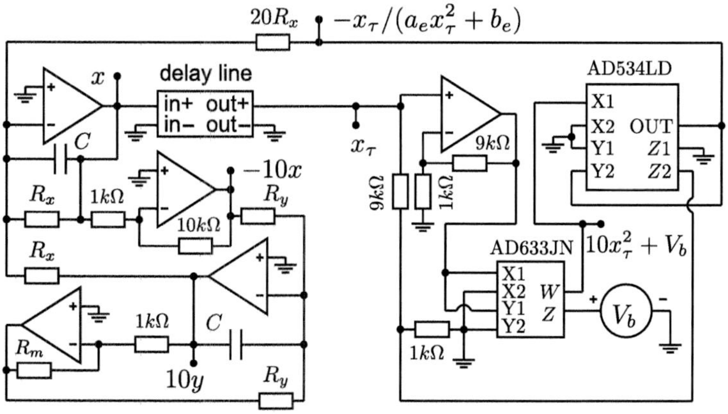

Fig. 5.29 Circuit diagram of the analog model for system (5.47). The used operational amplifiers are TL072CP. The circuit parameters are: $V_b = 10\,\text{mV}$, $C = 50\,\text{nF}$, $R_x = 1\,\text{k}\Omega$, $R_y = 20\,\text{k}\Omega$, $x_\tau = x(t-\tau)$, where $\tau = 47$ ms. The experimentally registered values of the nonlinearity parameters are $a_e = 10.92$, $b_e = 0.38$

Then the equations of the experimental setup can be transformed into dimensionless system (5.47) with $\varepsilon = \dfrac{0.1R_x}{R_y}$ by using substitution $t = t/\tau_0$ ($\tau_0 = R_yC = 10$ ms is the circuit's time constant) and new dynamical variables x/V_0 and y/V_0, where V_0 is the unity voltage, $V_0 = 1$ V.

References

1. H.M. Gibbs, *Optical Bistability: Controlling Light with Light* (Academic Press, New York, 1985)
2. P. Hänggi, P. Talkner, M. Borkovec, Rev. Mod. Phys. **62**(2), 251 (1990)
3. H. Risken, T. Frank, *The Fokker-Plank Equation: Methods of Solution and Application*. Springer Series in Synergetics (Springer, Berlin, 1996)
4. V. Semenov, S. Jalan, A. Zakharova, Chaos Solitons Fract. **173**, 113656 (2023)
5. V. Semenov, X. Porte, L. Larger, D. Brunner, Phys. Rev. B **108**(2), 024307 (2023)
6. V. Semenov, X. Porte, I. Abdulhalim, L. Larger, D. Brunner, Chaos **31**(12), 121104 (2021)
7. H. Kramers, Physica **7**(4), 284 (1940)
8. F. Schlögl, Z. Phys. **253**(2), 147 (1972)
9. A. Goldbeter, *Biochemical Oscillations and Cellular Rhythms* (Cambridge University Press, Cambridge, 1997)
10. G. Guidi, M.F. Carlier, A. Goldbeter, Biophys. J. **74**(3), 1229 (1998)
11. E. Ozbudak, M. Thattai, H. Lim, B. Shraiman, A. van Oudenaarden, Nature **427**(6976), 737 (2004)
12. A. Shilnikov, R.L. Calabrese, G. Cymbalyuk, Phys. Rev. E **71**(5), 056214 (2005)
13. A. Shilnikov, Nonlinear Dyn. **68**(3), 305 (2012)
14. W. Smits, O.P.Kuipers, J.W. Veening, Nat. Rev. Microbiol. **4**(4), 259 (2006)
15. R. Benzi, Nonlinear Process. Geophys. **17**(5), 431 (2010)

16. R.M. May, Nature **269**(5628), 471 (1977)
17. V. Guttal, C. Jayaprakash, Ecol. Model. **201**(3), 420 (2007)
18. R. Benzi, A. Sutera, A. Vulpiani, J. Phys. A Math. Gen. **14**(11), L453 (1981)
19. R. Benzi, G. Parisi, A. Sutera, A. Vulpiani, Tellus **34**(1), 10 (1982)
20. C. Nicolis, Tellus **34**(1), 1 (1982)
21. J. Freund, L. Schimansky-Geier, P. Hänggi, Chaos **13**(1), 225 (2003)
22. B. McNamara, K. Wiesenfeld, R. Roy, Phys. Rev. Lett. **60**(25), 2626 (1988)
23. A. Fioretti, L. Guidoni, R. Mannella, E. Arimondo, J. Stat. Phys. **70**(1), 403 (1993)
24. R. Bartussek, P. Hänggi, P. Jung, Phys. Rev. E **49**(5), 3930 (1994)
25. D. Luchinsky, R. Mannella, P. McClintock, N. Stocks, IEEE Trans. Circuits Syst. II Analog Digit. Signal Process. **46**(9), 1205 (1999)
26. D. Luchinsky, R. Mannella, P. McClintock, N. Stocks, IEEE Trans. Circuits Syst. II Analog Digit. Signal Process. **46**(9), 1215 (1999)
27. O. Calvo, D. Chialvo, Int. J. Bifur. Chaos **16**(3), 731 (2006)
28. A. Guderian, G. Dechert, K.P. Zeyer, F. Schneider, J. Phys. Chem. **100**(11), 4437 (1996)
29. A. Förster, M. Merget, F. Schneider, J. Phys. Chem. **100**(11), 4442 (1996)
30. W. Hohmann, J. Müller, F. Schneider, J. Phys. Chem. **100**(13), 5388 (1996)
31. A. Davtyan, M. Platkov, M. Gruebele, G. Papoian, ChemPhysChem **17**(9), 1305 (2016)
32. D. Leonard, L. Reichl, Phys. Rev. E **49**(2), 1734 (1994)
33. L. Yang, Z. Hou, H. Xin, J. Chem. Phys. **110**(7), 3591 (1999)
34. P. Hänggi, ChemPhysChem **3**(3), 285 (2002)
35. D. Hong, J. Martin, W. Saidel, Theor. Biol. Med. Model. **3**(1), 39 (2006)
36. M. McDonnell, D. Abbott, PLoS Comput. Biol. **5**(5), e1000348 (2009)
37. L. Stone, P. Sapatin, A. Huppert, C. Price, Geophys. Res. Lett. **25**(2), 175 (1998)
38. C. Nicolis, Sol. Phys. **74**(2), 473 (1981)
39. D. Valenti, F. Spagnolo, Phys. A Stat. Mech. Appl. **331**(3–4), 477 (2004)
40. J. Park, B. Kim, H. Park, Phys. Rev. E **104**(2), 024133 (2021)
41. A. Mubayi, C. Kribs, V. Arunachalam, C. Castillo-Chavez, in *Integrated Population Biology and Modeling, Part B*. ed. by S. Arni, C. Rao. Handbook of Statistics, vol. 40 (Elsevier, Amsterdam, 2019), pp. 157–193
42. A. Traulsen, T. Röhl, H. Schuster, Phys. Rev. Lett. **93**(2), 028701 (2004)
43. R. Zheng, K. Nakano, H. Hu, D. Su, M. Cartmell, J. Sound Vib. **333**(12), 2568 (2014)
44. J. Shu-Yao, Y. Fei, C. En, *OCEANS 2016 - Shanghai, 2016* (2016), pp. 1–5
45. Z. Qiao, Y. Lei, N. Li, Mech. Syst. Signal Process. **122**, 502 (2019)
46. T. Roy, P. Parmananda, Chaos **29**(9), 093121 (2019)
47. L. Gammaitoni, P. Hänggi, P. Jung, F. Marchesoni, Rev. Mod. Phys. **70**(1), 223 (1998)
48. V. Anishchenko, A. Neiman, F. Moss, L. Schimansky-Geier, Phys. Uspekhi **42**(1), 7 (1999)
49. M. Gosak, M. Perc, S. Kralj, Eur. Phys. J. B **80**(4), 519 (2011)
50. Y. Tang, W. Zou, J. Lu, J. Kurths, Phys. Rev. E **85**(4), 046207 (2012)
51. V. Semenov, A. Zakharova, Chaos **32**(12), 121106 (2022)
52. R. Benzi, A. Sutera, A. Vulpiani, J. Phys. A Math. Gen. **18**(12), 2239 (1985)
53. P. Jung, G. Mayer-Kress, Phys. Rev. Lett. **74**(11), 2130 (1995)
54. H. Wio, Phys. Rev. E **54**(4), R3075 (1996)
55. H. Wio, S. Bouzat, B. von Haeften, Phys. A Stat. Mech. Appl. **306**, 140 (2002)
56. Y.C. Lai, K. Park, L. Rajagopalan, Eur. Phys. J. B **69**(1), 65 (2009)
57. V. Semenov, A. Zakharova, Y. Maistrenko, E. Schöll, Europhys. Lett. **115**(1), 10005 (2016)
58. V. Semenov, A. Neiman, T. Vadivasova, V.S. Anishchenko, Phys. Rev. E **93**(5), 052210 (2016)
59. V. Semenov, A. Neiman, T. Vadivasova, V. Anishchenko, Izvestiya VUZ. Appl. Nonlinear Dyn. (in Russian) **24**(1), 5 (2016)
60. J. Nagumo, S. Arimoto, S. Yoshizawa, Proc. IRE **50**(10), 2061 (1962)
61. E.M. Izhikevich, R. FitzHugh, Scholarpedia **1**(9), 1349 (2006)
62. V. Semenov, Phys. Rev. E **95**(5), 052205 (2017)
63. V.V. Semenov, Chaos Solitons Fract. **116**, 358 (2018)
64. T. Bogatenko, V. Semenov, Phys. Lett. A **382**(37), 2645 (2018)

65. V.V. Semenov, A. Zakharova, Chaos **33**(8), 081101 (2023)
66. G. Kano, H. Iwasa, H. Tagaki, I. Teramoto, Electronics **48**(13), 105 (1975)
67. H. Poincaré, J. Math. Pures Appl. **7–8**, 375251 (1881–1882)
68. A. Andronov, A. Vitt, S. Khaikin, *Theory of Oscillators* (Pergamon Press, Oxford, 1966)
69. E. Hopf, Akad. Wiss. Leipzig **94**, 3 (1942)
70. C. Morawetz, J. Serrin, Y. Sinai (eds), *Selected Works of Eberhard Hopf with Commentaries* (American Mathematical Society, Providence, 2002)
71. J. Groszkowski, *Frequency of Self-Oscillations* (Pergamon Press, Oxford, 1964)
72. B. Hassard, N. Kazarinoff, Y.H. Wan, *Theory and Applications of Hopf Bifurcation* (Cambridge University Press, Cambridge, 1981)
73. A. Jenkins, Phys. Rep. **525**, 167 (2013)
74. L. Aguilar, I. Boiko, L. Fridman, R. Iriarte, *Self-Oscillations in Dynamic Systems* (Birkhäuser, Basel, 2015)
75. R. Field, M. Burger (eds), *Oscillations and Traveling Waves in Chemical Systems* (Wiley, Hoboken, 1985)
76. I. Epstein, J. Pojman, *An Introduction to Nonlinear Chemical Dynamics* (Oxford University Press, Oxford, 1998)
77. A. Mikhailov, G. Ertl (eds), *Engineering of Chemical Complexity* (World Scientific, Singapore, 2013)
78. A. Costa, G. Wadge, O. Melnik, Earth Planet. Sci. Lett. **337–338**, 39 (2012)
79. E. De Lauro, S. De Martino, M. Falanga, M. Palo, Nonlinear Process. Geophys. **18**, 925 (2011)
80. A. Berger, J. Imbrie, J. Hays, G. Kukla, B. Saltzman (eds), *Milankovitch and Climate* (Springer, Berlin, 1984)
81. M. Cane, M. Münnich, S. Zebiak, J. Atmos. Sci. **47**(13), 1562 (1990)
82. M. Münnich, M. Cane, S. Zebiak, J. Atmos. Sci. **48**(10), 1238 (1991)
83. E. Sel'kov, Eur. J. Biochem. **4**, 79 (1968)
84. W. Adey, A. Lawrence (eds), *Nonlinear Electrodynamics in Biological Systems* (Plenum Press, New York, 1984)
85. J. Walleczek (ed), *Self-Organized Biological Dynamics and Nonlinear Control* (Cambrdge University Press, Cambrdge, 2000)
86. E. Izhikevich, *Dynamical Systems in Neuroscience* (The MIT Press, Cambridge, 2007)
87. R. Goodwin, Econometrica **17**(2), 184 (1949)
88. Y. Mishchenko, PLoS One **9**(2), e87820 (2014)
89. S. Strogatz, *Nonlinear Dynamics and Chaos* (Perseus Books Publishing, New York, 1994)
90. S. Wiggins, *Introduction to Applied Nonlinear Dynamical Systems and Chaos* (Springer, Berlin, 2003)
91. V. Anishchenko, V. Astakhov, A. Neiman, T. Vadivasova, L. Schimansky-Geier, *Nonlinear Dynamics of Chaotic and Stochastic Systems* (Springer, Berlin, 2007)
92. W. Ebeling, H. Herzel, W. Richert, L. Schimansky-Geier, Z. Ange. Math. Mech. **66**(3), 141 (1986)
93. R. Lefever, J. Turner, Phys. Rev. Lett. **56**(16), 1631 (1986)
94. L. Fronzoni, R. Mannella, P. McClintock, F. Moss, Phys. Rev. A **36**(2), 834 (1987)
95. L. Arnold, *Random Dynamical Systems* (Springer, Berlin, 2003)
96. A. Zakharova, J. Kurths, T. Vadivasova, A. Koseska, PLoS One **6**(5), e19696 (2011)
97. V. Semenov, A. Listov, T. Vadivasova, Izvestiya VUZ. Appl. Nonlinear Dyn. (in Russian). **22**(5), 43 (2014)
98. A. Pikovsky, J. Kurths, Phys. Rev. Lett. **78**(5), 775 (1997)
99. B. Lindner, J. García-Ojalvo, A. Neiman, L. Schimansky-Geier, Phys. Rep. **392**, 321 (2004)
100. A. Zakharova, T. Vadivasova, V. Anishchenko, A. Koseska, J. Kurths, Phys. Rev. E **81**(1), 011106 (2010)
101. P.M. Geffert, A. Zakharova, A. Vüllings, W. Just, E. Schöll, Eur. Phys. J. B **87**, 291 (2014)
102. V. Semenov, A. Feoktistov, T. Vadivasova, E. Schöll, A. Zakharova, Chaos **25**(3), 033111 (2015)

103. M. Masoliver, N. Malik, E. Schöll, A. Zakharova, Chaos **27**(10), 101102 (2017)
104. A. Pikovsky, M. Rosenblum, J. Kurths, *Synchronization: A Universal Concept in Nonlinear Sciences* (Cambridge University Press, Cambridge, 2001)
105. D. Goldobin, A. Pikovsky, Phys. Rev. E **71**(4), 045201(R) (2005)
106. A. Zakharova, A. Feoktistov, T. Vadivasova, E. Schöll, Eur. Phys. J. Spec. Top. **222**(10), 2481 (2013)
107. B. Van der Pol, Radio Rev. **1**, 701 (1920)
108. B. Van der Pol, J. Van der Mark, Nature **120**, 363 (1920)
109. V. Semenov, K. Zakoretskii, T. Vadivasova, Russ. J. Nonlinear Dyn. **9**(3), 421 (2013)
110. V. Semenov, T. Vadivasova, V. Anishchenko, Tech. Phys. Lett. **39**(7), 632 (2013)
111. V. Anishchenko, T. Vadivasova, A. Feoktistov, V. Semenov, G. Strelkova, Experimental studies of noise effects in nonlinear oscillators, in *Nonlinear Dynamics and Complexity* (Springer, Berlin, 2014), pp. 261–290
112. I. Prigogine, P. Glansdorff, *Thermodynamic Theory of Structure, Stability and Fluctuations* (Wiley, Hoboken, 1971)
113. K. Zakoretskii, V. Semenov, T. Vadivasova, J. Commun. Technol. Electron. **61**(10), 1120 (2016)
114. H. Gang, T. Ditzinger, C. Ning, H. Haken, Phys. Rev. Lett. **71**(6), 807 (1993)
115. B. Lindner, L. Schimansky-Geier, Phys. Rev. E **60**(6), 7270 (1999)
116. L. DeVille, E. Vanden-Eijnden, C. Muratov, Phys. Rev. E **72**(3), 031105 (2005)
117. C. Muratov, E. Vanden-Eijnden, E. Weinan, Phys. D **210**(3–4), 227 (2005)
118. O. Ushakov, H.J. Wünsche, F. Henneberger, I. Khovanov, L. Schimansky-Geier, M. Zaks, Phys. Rev. Lett. **95**(12), 123903 (2005)
119. S.G. Lee, A. Neiman, S. Kim, Phys. Rev. E **57**(3), 3292 (1998)
120. T. Tateno, K. Pakdaman, Chaos **14**(3), 511 (2004)
121. B. Dmitriev, Y. Zharkov, S. Sadovnikov, V. Skorokhodov, A. Stepanov, Tech. Phys. Lett. **37**(11), 1082 (2011)
122. J. Hizanidis, A. Balanov, A. Amann, E. Schöll, Phys. Rev. Lett. **96**(24), 244104 (2006)
123. Y. Huang, H. Qin, W. Li, S. Lu, J. Dong, H. Grahn, Y. Zhang, Europhys. Lett. **105**(4), 47005 (2014)
124. G. Giacomelli, M. Giudici, S. Balle, J. Tredicce, Phys. Rev. Lett. **84**(15), 3298 (2000)
125. J. Avila, L.D.S. Cavalcante, J. Leite, Phys. Rev. Lett. **93**(14), 144101 (2004)
126. C. Otto, B. Lingnau, E. Schöll, K. Lüdge, Opt. Exp. **22**(11), 13288 (2014)
127. M. Arteaga, M. Valencia, M. Sciamanna, H. Thienpont, M. López-Amo, K. Panajotov, Phys. Rev. Lett. **99**(2), 023903 (2007)
128. F. Arecchi, R. Meucci, Eur. Phys. J. B **69**, 93 (2009)
129. L. Kabiraj, R. Steinert, A. Saurabh, C. Paschereit, Phys. Rev. E **92**(4), 042909 (2015)
130. P. Shaw, D. Saha, S. Ghost, M. Janaki, A. Iyengar, Chaos **25**(4), 043101 (2015)
131. K. Miyakawa, H. Isikawa, Phys. Rev. E **66**(4), 046204 (2002)
132. V. Beato, I. Sendiña-Nadal, I. Gerdes, H. Engel, Phil. Trans. R. Soc. A **366**(1864), 381 (2008)
133. D. Simakov, J. Pérez-Mercader, Sci. Rep. **3**, 2404 (2013)
134. S. Han, T. Yim, D. Postnov, O. Sosnovtseva, Phys. Rev. Lett. **83**(9), 1771 (1999)
135. M. Ciszak, O. Calvo, C. Masoller, C. Mirasso, R. Toral, Phys. Rev. Lett. **90**(20), 204102 (2003)
136. M. Ciszak, R. Toral, C. Mirasso, Mod. Phys. Lett. B **18**(23), 1135 (2004)
137. S. Astakhov, A. Feoktistov, V. Anishchenko, J. Kurths, Chaos **21**(4), 047513 (2011)
138. R. FitzHugh, Biophys. J. **1**(6), 445 (1961)
139. T. Kostova, R. Ravindran, M. Schonbek, Int. J. Bifur. Chaos **14**(3), 913 (2004)
140. S. Sehgal, A. Foulkes, Phys. Rev. E **102**(1), 012212 (2020)
141. V. Semenov, T. Vadivasova, E. Schöll, A. Zakharova, Izvestiya of Saratov University (in Russian). Physics **15**(3), 43 (2015)
142. E. Lorenz, J. Atmos. Sci. **20**(2), 130 (1963)
143. O. Rössler, Phys. Lett. A **57**(5), 397 (1976)
144. T. Matsumoto, IEEE Trans. Circuits Syst. **31**(12), 1055 (1984)

145. T. Matsumoto, L. Chua, M. Komuro, IEEE Trans. Circuits Syst. **32**(8), 797 (1985)
146. V.S. Anishchenko, V. Astakhov, Radiotekh. Elektron. (in Russian) **28**, 1109 (1983)
147. V. Anishchenko, *Dynamical Chaos - Models and Experiments* (World Scientific, Singapore, 1995)
148. K. Ikeda, Opt. Commun. **30**(2), 257 (1979)
149. K. Ikeda, H. Daido, O. Akimoto, Phys. Rev. Lett. **45**(9), 709 (1980)
150. M. Mackey, L. Glass, Science **197**(4300), 287 (1977)
151. K. Lehnertz, C.E. Elger, J. Arnhold, P. Grassberger, *Chaos in Brain?* (World Scientific, Singapore, 2000)
152. M. Rabinovich, H. Abarbanel, Neuroscience **87**(1), 5 (1998)
153. W. Freeman, Int. J. Bifur. Chaos **2**(3), 451 (1992)
154. R. Field, L. Györgyi (eds), *Chaos in Chemistry and Biochemistry* (World Scientific, Singapore, 1993)
155. S. Scott, *Oscillations, Waves, and Chaos in Chemical Kinetics* (Oxford University Press, Oxford, 1994)
156. C. Pahl-Wostl, *The Dynamic Nature of Ecosystems: Chaos and Order Entwined* (Wiley, Hoboken, 1995)
157. J. Cushing, R. Costantino, R. Dennis, B. Desharnais, S. Henson, *Chaos in Ecology: Experimental Nonlinear Dynamics* (Academic Press, Cambridge, 2002)
158. A. Medvinskii, S. Petrovskii, I. Tikhonova, D. Tikhonov, B. Li, E. Venturino, H. Malchow, G. Ivanitskii, Phys. Uspekhi **45**(1), 27 (2002)
159. P. Mitkowski, *Mathematical Structures of Ergodicity and Chaos in Population Dynamics* (Springer, Berlin, 2021)
160. S. Gokhale, A. Conwill, T. Ranjan, J. Gore, Nat. Commun. **9**(1), 5273 (2018)
161. E. Hopf, M. Derstine, H.M. Gibbs, M. Rushford, *Chaos in Optics* (Springer, Berlin, 1984), pp. 67–79
162. F. Arecchi, R. Harrison (eds), *Instabilities and Chaos in Quantum Optics*. Springer Series in Synergetics, vol. 34 (Springer, Berlin, 1987)
163. J. Ohtsudo, *Semiconductor Lasers* (Springer, Berlin, 2013)
164. L. Fan, X. Yan, H. Wang, L. Wang, Sci. Adv. **7**(3), eabc8448 (2021)
165. M. Wyk, W.H. Steeb, *Chaos in Electronics* (Springer, Berlin, 1997)
166. K. Aoki, *Nonlinear Dynamics and Chaos in Semiconductors* (CRC Press, Boca Raton, 2000)
167. E. Schöll, *Nonlinear Spatio-Temporal Dynamics and Chaos in Semiconductors* (Cambridge University Press, Cambridge, 2001)
168. G. Chen, T. Ueta (eds), *Chaos in Circuits and Systems* (World Scientific, Singapore, 2002)
169. W. Horton, Y.H. Ichikawa, *Chaos and Structures in Nonlinear Plasmas* (World Scientific, Singapore, 1996)
170. Y. Elskens, D. Escande, *Microscopic Dynamics of Plasmas and Chaos* (CRC Press, Boca raton, 2001)
171. D. Turcotte, *Fractals and Chaos in Geology and Geophysics* (Cambrdge University Press, Cambrdge, 1997)
172. W. Baumol, J. Benhabib, J. Econ. Perspect. **3**(1), 77 (1989)
173. R. Day, Quart. J. Econ. **98**(2), 201 (1983)
174. H.W. Lorenz, *Nonlinear Dynamical Economics and Chaotic Motion* (Springer, Berlin, 1989)
175. R. Goodwin, *Chaotic Economic Dynamics* (Oxford University Press, Oxford, 1990)
176. J. Farmer, Phys. D **4**(3), 366 (1982)
177. M. Lakshmanan, D. Senthilkumar, *Dynamics of Nonlinear Time-Delay Systems* (Springer, Berlin, 2011)
178. D. Biswas, T. Banerjee, *Time-Delayed Chaotic Dynamical Systems* (Springer, Berlin, 2018)
179. L. Schimansky-Geier, H. Herzel, J. Stat. Phys. **70**(1), 141 (1993)
180. I. Percival, Proc. R. Soc. Lond. A **413**(1844), 131 (1987)
181. H. Dankowicz, *Chaotic Dynamics in Hamiltonian Systems* (World Scientific, Singapore, 1997)
182. G. Leonov, N. Kuznetsov, Int. J. Bifur. Chaos **23**(1), 1330002 (2013)

183. V.T. Pham, C. Volos, S. Jafari, T. Kapitaniak, Nonlinear Dyn. **87**(3), 2001 (2017)
184. X. Wang, N. Kuznetsov, G. Chen (eds), *Chaotic Systems with Multistability and Hidden Attractors* (Springer, Berlin, 2021)
185. K. Kaneko, Phys. D **34**(1–2), 1 (1989)
186. V. Malyaev, V. Semenov, T. Vadivasova, Izvestiya VUZ Appl. Nonlinear Dyn. (in Russian) **20**(3), 17 (2012)
187. V. Anishchenko, T. Vadivasova, G. Strelkova, *Deterministic Nonlinear Systems* (Springer, Berlin, 2014)
188. V. Semenov, Izvestiya VUZ Appl. Nonlinear Dyn. (in Russian) **21**(3), 43 (2013)
189. L. Chua, IEEE Trans. Circuit Theory **CT-18**(5), 507 (1971)
190. L. Chua, S. Kang, Proc. IEEE **64**(2), 209 (1976)
191. M. Di Ventra, Y. Pershin, L. Chua, Proc. IEEE **97**(10), 1717 (2009)
192. L. Chua, Appl. Phys. A **102**(4), 765 (2011)
193. L. Chua, Nanotechnology **24**(38), 383001 (2013)
194. G. Gandhi, V. Aggarwal, L. Chua, IEEE Circuits Syst. Mag. **13**(2), 8 (2013)
195. D. Strukov, G. Snider, D. Stewart, R. Williams, Nature **453**(7191), 80 (2008)
196. F. Argall, Solid-State Electron. **11**(5), 535 (1968)
197. A. Beck, J. Bednorz, C. Gerber, C. Rossel, D. Widmer, Appl. Phys. Lett. **77**(1), 139 (2000)
198. A. Michaylov, A. Belov, D. Guseinov, D. Korolev, I. Antonov, D. Efimovykh, S. Tikhov, A. Kasatkin, O. Gorshkov, D. Telebaum, A. Bobrov, N. Malekhonova, D. Pavlov, E. Gryaznov, A. Yatmanov, Mater. Sci. Eng. B **194**, 48 (2015)
199. G. Dearnaley, A. Stoneham, D. Morgan, Rep. Prog. Phys. **33**(3), 1129 (1970)
200. A. Sawa, Mater. Today **11**(6), 28 (2008)
201. Y. Yang, P. Sheridan, W. Lu, Appl. Phys. Lett. **100**(20), 203112 (2012)
202. S. Kim, S. Choi, W. Lu, ACS Nano **8**(3), 2369 (2014)
203. J. Strachan, A. Torrezan, F. Miao, M. Pickett, J. Yang, W. Yi, G. Medeiros-Ribeiro, R. Williams, IEEE Trans. Electron. Devices **60**(7), 2194 (2013)
204. S. Wu, X. Luo, H. Peng, W. Lin, J. Ding, A. David, B. Wang, G. Tendeloo, J. Wang, T. Wu, Phys. Rev. X **3**(4), 041027 (2013)
205. A. Mehonic, S. Cueff, M. Wojdak, S. Hudziak, O. Jambois, C. Labbé, B. Garrido, R. Rizk, A. Kenyon, J. Appl. Phys. **111**(7), 074507 (2012)
206. H. Jeong, J. Kim, J. Kim, J. Hwang, J. Kim, J. Lee, T. Yoon, B. Cho, S. Kim, R. Ruoff, S. Choi, Nano Lett. **10**(11), 4381 (2010)
207. J. Lee, C. Du, K. Sun, E. Kioupakis, W. Lu, ACS Nano **10**(3), 3571 (2016)
208. J. Krieger, S. Spitzer, *Proceedings of the 2004 Non-Volatile Memory Technology Symposium* (IEEE, Piscataway, 2004), pp. 121–124
209. G. Liu, Y. Chen, C. Wang, W. Zhang, R.W. Li, L. Wang, Mater. Horizons **1**(5), 489 (2014)
210. V. Demin, V. Erokhin, A. Emelyanov, S. Battistoni, G. Baldi, S. Iannotta, P. Kashkarov, M. Kovalchuk, Organ. Electron. **25**, 16 (2015)
211. T. Berzina, A. Smerieri, M. Bernabò, A. Pucci, G. Ruggeri, V. Erokhin, M. Fontana, J. Appl. Phys. **105**(12), 124515 (2009)
212. S. Erokhina, V. Sorokin, V. Erokhin, Electron. Mater. Lett. **11**(5), 801 (2015)
213. J. Moll, Y. Tarui, IEEE Trans. Electron. Devices **10**(5), 338 (1963)
214. A. Chanthbouala, V. Garcia, R. Cherifi, K. Bouzehouane, S. Fusil, X. Moya, S. Xavier, H. Yamada, C. Deranlot, N. Mathur, M. Bibes, A. Barthélémy, J. Grollier, Nat. Mater. **11**(10), 860 (2012)
215. F. Buot, A. Rajagopal, J. Appl. Phys. **76**(9), 5552 (1994)
216. Y. Yilmaz, P. Mazumder, *Proceedings of the 2012 Nanotechnology Conference* (IEEE, Piscataway, 2012), pp. 1–4
217. Y. Yilmaz, P. Mazumder, IEEE Trans. Nanotechnol. **12**(6), 879 (2013)
218. X. Wang, Y. Chen, H. Xi, H. Li, D. Dimitrov, IEEE Electron. Devices Lett. **30**(3), 294 (2009)
219. A. Chanthbouala, R. Matsumoto, J. Grollier, V. Cros, A. Anane, A. Fert, A. Khvalkovskiy, K. Zvezdin, K. Nishimura, Y. Nagamine, H. Maehara, K. Tsunekawa, A. Fukushima, S. Yuasa, Nat. Phys. **7**(8), 626 (2011)

220. Y. Pershin, M. Di Ventra, Phys. Rev. B **77**(7), 073301 (2008)
221. B. Muthuswamy, International Journal of Bifurcation and Chaos **20**(5), 1335 (2010)
222. B. Bao, J. Yu, F. Hu, Z. Liu, International Journal of Bifurcation and Chaos **24**(11), 1450143 (2014)
223. R.C. Johnson, EE Times **1538**, 30 (2008)
224. J. Tour, T. He, Nature **453**, 42 (2008)
225. M. Di Ventra, Y. Pershin, Nat. Phys. **9**(4), 200 (2013)
226. J. Yang, D. Strukov, D. Stewart, Nat. Nanotechnol. **8**(1), 13 (2013)
227. Y. Pershin, M. Di Ventra, IEEE Trans. Circuits Syst. **57**(8), 1857 (2010)
228. S. Shin, K. Kim, S. Kang, IEEE Trans. Nanotechnol. **10**(2), 266 (2011)
229. Z. Chew, L. Li, Electron. Lett. **48**(25), 1610 (2012)
230. L. Gao, F. Merrikh-Bayat, F. Alibart, X. Guo, B. Hoskins, K. Cheng, D. Strukov, *Proceedings of the 2013 IEEE/ACM International Symposium on Nanoscale Architectures (NANOARCH)* (2013), pp. 19–22
231. A. Ascoli, R. Tetzlaff, F. Corinto, M. Mirchev, M. Gilli, *Proceedings of the 2013 14th Latin American Test Workshop - LATW* (2013), pp. 1–6
232. H. Ho-Ching Iu, A. Fitch, *Development of Memristor Based Circuits*. World Scientific Series on Nonlinear Science, Series A, vol. 82 (World Scientific, Singapore, 2012)
233. A. Radwan, M. Fouda, *On the Mathematical Modeling of Memristor, Memcapacitor, and Meminductor*. Studies in Systems, Decision and Control, vol. 26 (Springer, Berlin, 2015)
234. R. Tetzlaff (ed), *Memristors and Memristive Systems* (Springer, New York, 2014)
235. A. Adamatzky, L. Chua (eds), *Memristor Networks* (Springer, Berlin, 2014)
236. I. Vourkas, G. Sirakoulis, *Memristor-Based Nanoelectronic Computing Circuit and Architectures*. Emergence, Complexity and Computation, vol. 19 (Springer, Berlin, 2016)
237. R. Kozma, R. Pino, G. Pazienza (eds), *Advanced in Neuromorphic Memristor Science and Applications*. Springer Series in Cognitive and Neural Systems, vol. 4 (Springer, Dordrecht, 2012)
238. S. Jo, T. Chang, I. Ebong, B. Bhadviya, P. Mazumder, W. Lu, Nano Lett. **10**(4), 1297 (2010)
239. Y. Pershin, M. Di Ventra, Neural Netw. **23**(7), 881 (2010)
240. A. Williamson, L. Schurmann, L. Hiller, F. Klefenz, I. Hoerselmann, P. Husar, A. Schober, Nanoscale **5**(16), 7297 (2013)
241. Y. Li, Y. Zhong, L. Xu, J. Zhang, H. Xu, X. Sun, X. Miao, Sci. Rep. **3**, 1619 (2013)
242. A. Serb, J. Bill, A. Khiat, R. Berdan, R. Legenstein, T. Prodromakis, Nat. Commun. **7**, 12611 (2016)
243. P. Sheridan, C. Du, W. Lu, IEEE Trans. Neural Netw. Learn. Syst. **27**(11), 2327 (2016)
244. S. Gaba, P. Sheridan, J. Zhou, S. Choi, W. Lu, Nanoscale **5**(13), 5872 (2013)
245. A. Emelyanov, D. Lapkin, V. Demin, V. Erokhin, S. Battistoni, G. Baldi, A. Dimonte, A. Korovin, S. Iannotta, P. Kashkarov, M. Kovalchuk, AIP Adv. **6**(11), 111301 (2016)
246. M. Ziegler, R. Soni, T. Patelczyk, M. Ignatov, T. Bartsch, P. Meuffels, H. Kohlstedt, Adv. Function. Mater. **22**(13), 2744 (2012)
247. S. Vaidyanathan, C. Volos, *Advances in Memristors, Memristive Devices and Systems* (Springer, Berlin, 2017)
248. D. Ielmini, R. Waser (eds), *Resistive Switching: From Fundamentals of Nanoionic Redox Processes to Memristive Device Applications* (Wiley, Hoboken, 2016)
249. Z. Wang, S. Joshi, S. Savel'ev, H. Jiang, M. R., P. Lin, M. Hu, N. Ge, J. Strachan, Z. Li, Q. Wu, M. Barnel, G.L. Li, H. Xin, R. Williams, Q. Xia, J. Yang, Nat. Mater. **16**(1), 101 (2017)
250. G. Adam, B. Hoskins, M. Prezioso, F. Merrikh-Bayat, B. Chakrabarti, D. Strukov, IEEE Trans. Electron. Devices **64**(1), 312 (2017)
251. S. Hamdioui, S. Kvatinsky, G. Cauwenberghs, L. Xie, N. Wald, S. Joshi, H. Elsayed, H. Corporaal, K. Bertels, *2017 Design, Automation and Test in Europe Conference and Exhibition (27–31 March 2017)* (IEEE, Piscataway, 2017), pp. 722–731
252. P. Sheridan, F. Cai, C. Du, W. Ma, Z. Zhang, W.D. Lu, Nat. Nanotechnol. **12**, 784 (2017)
253. T. Zhang, M. Yin, Y. Cai, Y. Yang, R. Huang, Nano Futures **1**(1), 015003 (2017)

254. M. Itoh, L. Chua, Int. J. Bifur. Chaos **18**(11), 3183 (2008)
255. A. Buscarino, L. Fortuna, M. Frasca, L. Gambuzza, Chaos **22**(2), 023136 (2012)
256. A. Buscarino, L. Fortuna, M. Frasca, L. Gambuzza, Int. J. Bifur. Chaos **23**(5), 1330015 (2013)
257. V.T. Pham, A. Buscarino, L. Fortuna, M. Frasca, Int. J. Bifur. Chaos **23**(4), 1350073 (2013)
258. L. Gambuzza, L. Fortuna, M. Frasca, E. Gale, Int. J. Bifur. Chaos **25**(8), 1550101 (2015)
259. M. Itoh, L. Chua, Int. J. Bifur. Chaos **21**(9), 2395 (2011)
260. M. Frasca, L. Gambuzza, A. Buscarino, L. Fortuna, Phys. Status Solidi **12**(1–2), 206 (2014)
261. C. Volos, I. Kyprianidis, I. Stouboulos, J. Muñoz-Pacherco, V.T. Pham, J. Eng. Sci. Technol. Rev. **8**(2), 44 (2015)
262. C. Volos, V.T. Pham, S. Vaidyanathan, I. Kyprianidis, I. Stouboulos, The case of bidirectionally coupled nonlinear circuits via a memristor, in *Advances and Applications in Nonlinear Control Systems*. Studies in Computational Intelligence (Springer, Berlin, 2016)
263. M. Frasca, L. Gambuzza, A. Buscarino, L. Fortuna, *Advanced in Neural Networks: Computational and Theoretical Issues* (Springer, Berlin, 2015); *Smart Innovation, Systems and Technologies*, vol. 37, chap. Memristor Based Adaptive Coupling for Synchronization of Two Rössler Systems, pp. 295–300
264. M. Ignatov, M. Hansen, M. Ziegler, H. Kohlstedt, Appl. Phys. Lett. **108**(8), 084105 (2016)
265. I. Korneev, O. Shabalina, V. Semenov, T. Vadivasova, Izvestiya VUZ. Appl. Nonlinear Dyn. (in Russian) **26**(2), 24 (2018)
266. I. Korneev, V. Semenov, T. Vadivasova, Int. J. Bifur. Chaos **30**(7), 2050096 (2020)
267. I. Korneev, V. Semenov, A. Slepnev, T. Vadivasova, Chaos Solitons Fract. **142**, 110459 (2021)
268. V.T. Pham, A. Buscarino, L. Fortuna, M. Frasca, Int. J. Bifur. Chaos **22**(8), 1230027 (2012)
269. L. Gambuzza, A. Buscarino, L. Fortuna, M. Frasca, IEEE Trans. Circuits Syst. **62**(4), 1175 (2015)
270. A. Buscarino, C. Corradino, L. Fortuna, M. Frasca, L.O. Chua, Turing Patterns in Memristive Cellular Nonlinear Networks. IEEE Trans. Circuits Syst. I Regular Papers **63**(8), 1222–1230 (2016). https://doi.org/10.1109/TCSI.2016.2564738
271. H. Zhao, L. Li, H. Peng, J. Kurths, J. Xiao, Y. Yang, Eur. Phys. J. B **88**(5), 1 (2015)
272. I. Korneev, A. Slepnev, V. Semenov, T. Vadivasova, Izvestiya of Saratov University. Phys (in Russian) **20**(3), 210 (2020)
273. I. Korneev, V. Semenov, A. Slepnev, T. Vadivasova, Chaos Solitons Fract **147**, 110923 (2021)
274. I. Korneev, I. Ramazanov, V. Semenov, A. Slepnev, T. Vadivasova, Front. Phys. **10**, 886476 (2022)
275. I. Korneev, A. Slepnev, V. Semenov, T. Vadivasova, Izvestiya VUZ. Appl. Nonlinear Dyn. (in Russian) **28**(3), 324 (2020)
276. I.A. Korneev, I.R. Ramazanov, A.V. Slepnev, T.E. Vadivasova, V.V. Semenov, Travelling waves in an ensemble of excitable oscillators: the interplay of memristive coupling and noise. Chaos **34**(12), 123110 (2024). https://doi.org/10.1063/5.0215152
277. I. Korneev, I. Ramazanov, V. Semenov, A. Slepnev, T. Vadivasova, Commun. Nonlinear Sci. Numer. Simul. **135**, 108072 (2024)
278. A. Stotland, M. Di Ventra, Phys. Rev. E **85**(1), 011116 (2012)
279. V.T. Pham, C. Volos, S. Vaidyanathan, T. Le, V. Vu, J. Eng. Sci. Technol. Rev. **8**(2), 205 (2015)
280. M. Chen, J. Yu, B. Bao, Electron. Lett. **51**(6), 462 (2015)
281. M. Chen, M. Li, Q. Yu, B. Bao, Q. Xu, J. Wang, Nonlinear Dyn. **81**(1), 215 (2015)
282. I. Korneev, A. Slepnev, T. Vadivasova, V. Semenov, Chaos **31**(7), 073102 (2021)
283. E. Linn, A. Siemon, R. Waser, S. Menzel, IEEE Trans. Circuits Syst. I Regul. Pap. **61**(8), 2402 (2014)
284. J. Singh, B. Raj, J. Comput. Electron. **18**(2), 640 (2019)
285. A. Ascoli, V. Senger, R. Tetzlaff, IEEE Circuits Syst. Mag. **13**(2), 89 (2013)
286. T. Chang, S. Jo, K. Kim, P. Sheridan, S. Gaba, W. Lu, Appl. Phys. A **102**(4), 857 (2011)
287. D. Guseinov, I. Matyushkin, N. Chernyaev, A. Mikhailov, Y. Pershin, Chaos Solitons Fract. **144**, 110699 (2021)
288. L. Chen, C. Li, T. Huang, Y. Chen, S. Wen, J. Qi, Phys. Lett. A **377**(45–48), 3260 (2013)

289. E. Zhou, L. Fang, B. Yang, Phys. Lett. A **383**(11), 942 (2019)
290. V. Semenov, Int. J. Electron. Lett. **12**(1), 59 (2024)
291. D. Yu, Y. Liang, H. Chen, H. Iu, IEEE Trans. Circuits Syst. II Exp. Briefs **60**(4), 207 (2013)
292. H. Bao, A. Hu, W. Liu, B. Bao, IEEE Trans. Neural Netw. Learn. Syst. **31**(2), 502 (2019)
293. A. Akther, Y. Ushakov, A.G. Balanov, S.E. Savel'ev, Chaos **31**(7), 073111 (2021)
294. V. Semenov, I. Korneev, P. Arinushkin, G. Strelkova, T. Vadivasova, V. Anishchenko, Eur. Phys. J. Spec. Top. **224**(8), 1553 (2015)
295. B. Fiedler, S. Liebscher, J. Alexander, J. Differ. Equ. **167**(1), 16 (2000)
296. B. Fiedler, S. Liebscher, SIAM J. Math. Anal. **31**(6), 1396 (2000)
297. B. Fiedler, S. Liebscher, J. Alexander, Int. J. Bifur. Chaos **10**(7), 1613 (2000)
298. S. Liebscher, *Bifurcation without Parameters*. Lectures Notes in Mathematics, vol. 2117 (Springer, Berlin, 2015)
299. S. Jafari, J. Sprott, V.T. Pham, C. Volos, C. Li, Nonlinear Dyn. **86**(2), 1349 (2016)
300. T. Gotthans, J. Petržela, Nonlinear Dyn. **81**(3), 1143 (2015)
301. T. Gotthans, J. Sprott, J. Petrzela, Int. J. Bifur. Chaos **26**(8), 1650137 (2016)
302. R. Riaza, SIAM J. Appl. Math. **72**(3), 877 (2012)
303. R. Riaza, SIAM J. APPL. Math. **78**(1), 395 (2018)
304. F. Corinto, M. Forti, L. Chua, *Nonlinear Circuits and Systems with Memristors* (Springer, Berlin, 2020)
305. M. Messias, C. Nespoli, V. Botta, Int. J. Bifur. Chaos **20**(2), 437 (2010)
306. V. Botta, C. Néspoli, M. Messias, TEMA Tend. Mat. Apl. Comput. **12**(2), 91 (2011)
307. I. Korneev, T. Vadivasova, V. Semenov, Nonlinear Dyn. **89**(4), 2829 (2017)
308. I. Korneev, V. Semenov, Chaos **27**(8), 081104 (2017)
309. I. Korneev, A. Slepnev, V. Semenov, T. Vadivasova, EPJ-ST **229**(12), 2215 (2020)
310. I. Korneev, I. Ramazanov, A. Slepnev, T. Vadivasova, V. Semenov, Chaos **33**(9), 091102 (2023)
311. I. Korneev, A. Slepnev, A. Zakharova, T. Vadivasova, V. Semenov, Nonlinear Dyn. **111**(2), 1235 (2023)
312. F.T. Arecchi, G. Giacomelli, A. Lapucci, R. Meucci, Phys. Rev. A **45**(7), R4225 (1992)
313. G. Giacomelli, A. Politi, Phys. Rev. Lett. **76**(15), 2686 (1996)
314. G. Giacomelli, F. Marino, M.A. Zaks, S. Yanchuk, Europhys. Lett. **99**(5), 58005 (2012)
315. L. Larger, B. Penkovsky, Y. Maistrenko, Nat. Commun. **6**(1), 7752 (2015)
316. L. Larger, B. Penkovsky, Y. Maistrenko, Phys. Rev. Lett. **111**(5), 054103 (2013)
317. D. Brunner, B. Penkovsky, R. Levchenko, E. Schöll, L. Larger, Y. Maistrenko, Chaos Interdiscip. J. Nonlinear Sci. **28**(10), 103106 (2018)
318. V.V. Semenov, Y.L. Maistrenko, Chaos Interdiscip. J. Nonlinear Sci. **28**(10), 101103 (2018)
319. S. Yanchuk, S. Ruschel, J. Sieber, M. Wolfrum, Phys. Rev. Lett. **123**(5), 053901 (2019)
320. S. Yanchuk, G. Giacomelli, J. Phys. A Math. Theor. **50**(10), 103001 (2017)
321. S. Fauve, O. Thual, Phys. Rev. Lett. **64**(3), 282 (1990)
322. B.S. Kerner, V.V. Osipov, *Autosolitons: A New Approach to Problems of Self-Organization and Turbulence* (Springer, Berlin, 1994)
323. C.I. Christov, M.G. Velarde, Phys. D **86**(1–2), 323 (1995)
324. F. Gustave, L. Columbo, G. Tissoni, M. Brambilla, F. Prati, B. Kelleher, B. Tykalewicz, S. Barland, Phys. Rev. Lett. **115**(4), 043902 (2015)
325. N. Akhmediev, A. Ankiewicz (eds), *Dissipative Solitons: From Optics to Biology and Medicine* (Springer, Berlin, 2008)
326. H.G. Purwins, H.U. Bödeker, S. Amiranashvili, Adv. Phys. **59**(5), 485 (2010)
327. A.W. Liehr, *Dissipative Solitons in Reaction Diffusion Systems* (Springer, Berlin, 2013)
328. A.D. Boardman, A.P. Sukhorukov (eds), *Soliton-Driven Photonics* (Springer, Berlin, 2001)
329. F. Haudin, R. Rojas, U. Bortolozzo, S. Residori, M. Clerc, Phys. Rev. Lett. **107**(26), 264101 (2011)
330. P. Grelu, N. Akhmediev, Nat. Photon. **6**, 84 (2012)
331. N. Verschueren, U. Bortolozzo, M. Clerc, S. Residori, Phys. Rev. Lett. **110**(10), 104101 (2013)

332. B. Garbin, J. Javaloyes, G. Tissoni, S. Barland, Nat. Commun. **6**, 5915 (2015)
333. M. Marconi, J. Javaloyes, S. Barland, S. Balle, M. Giudici, Nat. Photon. **9**, 450 (2015)
334. J. Javaloyes, M. Marconi, M. Giudici, Phys. Rev. Lett. **119**(3), 033904 (2017)
335. C.H. Wong, Y. Tserkovnyak, Phys. Rev. B **81**(6), 060404 (2010)
336. S.V. Grishin, E.N. Beginin, M.A. Morozova, Y.P. Sharaevskii, S.A. Nikitov, J. Appl. Phys. **115**(5), 053908 (2014)
337. A.V. Tur, A.V. Chechkin, V.V. Yanovsky, Phys. Fluids B **4**(11), 3515 (1992)
338. S. Ghost, A. Adak, M. Khan, Phys. Plasmas **21**(1), 012303 (2014)
339. S. Sultana, I. Kourakis, Phys. Plasmas **22**(10), 102302 (2015)
340. B. Lautrup, R. Appali, A.D. Jackson, T. Heimburg, Eur. Phys. J. E **34**(6), 57 (2011)
341. N. Akhmediev, J.M. Soto-Crespo, H.R. Brand, Phys. Lett. A **377**(13), 968 (2013)
342. S. Kai, H. Miike, Phys. A **204**(1–4), 346 (1994)
343. S. Kai, T. Ariyoshi, S. Inenaga, H. Miike, Phys. D **84**(1–2), 269 (1995)
344. A. Hasegawa, Y. Kodama, *Solitons in Optical Communications* (Clarendon Press, Oxford, 1995)
345. P. Marin-Palomo, J.N. Kemal, M. Karpov, A. Kordts, J. Pfeifle, M.H.P. Pfeiffer, P. Trocha, S. Wolf, V. Brasch, M.H. Anderson, R. Rosenberger, K. Vijayan, W. Freude, T.J. Kippenberg, C. Koos, Nature **546**, 274 (2017)
346. A. Adamatzky (ed), *Collision-Based Computing* (Springer, Berlin, 2002)
347. L. Larger, M. Soriano, D. Brunner, L. Appeltant, J. Gutierrez, L. Pesquera, C. Mirasso, I. Fisher, Opt. Exp. **20**(3), 3241 (2012)
348. Y. Paquot, F. Duport, A. Smerieri, J. Dambre, B. Schrauwen, M. Haelterman, S. Massar, Sci. Rep. **2**, 287 (2012)
349. B. Romeira, R. Avó, J.M.L. Figueiredo, S. Barland, J. Javaloyes, Sci. Rep. **6**, 19510 (2016)
350. A. Vladimirov, D. Turaev, Phys. Rev. A **72**(3), 033808 (2005)
351. M. Nizette, D. Rachinskii, A. Vladimirov, M. Wolfrum, Phys. D **218**(1), 95 (2006)
352. D. Puzyrev, A. Vladimirov, S. Gurevich, S. Yanchuk, Phys. Rev. A **93**(4), 041801 (2016)
353. V. Semenov, A. Bukh, N. Semenova, Chaos Solitons Fract. **172**, 113524 (2023)
354. Y. Kivshar, G. Agrawal, *Optical Solitons: From Fibers to Photonic Crystals* (Academic Press, Cambridge, 2003)

GPSR Compliance

The European Union's (EU) General Product Safety Regulation (GPSR) is a set of rules that requires consumer products to be safe and our obligations to ensure this.

If you have any concerns about our products, you can contact us on ProductSafety@springernature.com

In case Publisher is established outside the EU, the EU authorized representative is:

Springer Nature Customer Service Center GmbH
Europaplatz 3
69115 Heidelberg, Germany

Batch number: 10756341

Printed by Printforce, the Netherlands